'Clearly, the Islamic State's particular brand of brutality derives from the deeply dysfunctional and violent politics of the Arab states it has fought in Iraq, Syria, Libya and Egypt. The special irony, however, is that it proved far more adept than these foes at demonstrating strategic flexibility and at systematically instilling tactical autonomy, initiative, and unceasing offensive action among its fighters down to the lowest rank. Ashour's excellent, methodical dissection of this fluid interaction of strategy and tactics provides a com___ ___ explanation of how Islamic State fighters repe___ ___ ____ ____ lured against overwhelming odds.'
Professor Yezid Sayigh, ___ **and author of** *Armed Struggle an___*

'This meticulous examination a___ ___ ___ ___ ___ w the Islamic State has conducted its military c___ ___ across the Middle East and North Africa helps significantly to explain how it endured against such overwhelming odds. This book will undoubtedly become a standard work for anyone interested in understanding how a small terrorist group can morph into a proto-state.'
Richard Barrett OBE, CMG, Former Head of the United Nations ISIL (Da'esh), Al Qaida and Taliban Monitoring Team

'Omar Ashour delivers an impressively detailed study of ISIS war tactics and strategy in Iraq, Syria, Libya and Egypt. He therefore helps us understand not only how the jihadi networks could sustain repeated assaults by formidable enemies, but also how they could survive their current "defeat" and soon strike back.'
Professor Jean-Pierre Filiu, Paris Institute of Political Studies (Sciences Po) and author of *From Deep State to Islamic State*

'How ISIS Fights: Military Tactics in Iraq, Syria, Libya and Egypt greatly enhances our understanding of the group by focusing on the hitherto neglected dimension of its remarkable – and fortunately, ephemeral – tactical military successes. In a relatively short span of time, ISIS emerged as a major regional and then global threat. Omar Ashour masterfully explains ISIS's ways of war on battlefields that extended from Sirte to Mosul and from Raqqa to Sharm el-Sheikh. This an important book that fills a conspicuous gap in the literature.'
Professor Bruce Hoffman, Georgetown University and author of *Inside Terrorism*

To the beloved ones to whom I couldn't bid a farewell

How ISIS Fights

Military Tactics in Iraq, Syria, Libya and Egypt

Omar Ashour

EDINBURGH
University Press

Edinburgh University Press is one of the leading university presses in the UK. We publish academic books and journals in our selected subject areas across the humanities and social sciences, combining cutting-edge scholarship with high editorial and production values to produce academic works of lasting importance. For more information visit our website: edinburghuniversitypress.com

Edinburgh University Press Ltd
The Tun – Holyrood Road
12 (2f) Jackson's Entry
Edinburgh EH8 8PJ

Typeset in 10.5/12.5 Bembo by
IDSUK (DataConnection) Ltd, and

A CIP record for this book is available from the British Library

ISBN 978 1 4744 3821 6 (hardback)
ISBN 978 1 4744 3822 3 (paperback)
ISBN 978 1 4744 3823 0 (webready PDF)
ISBN 978 1 4744 3824 7 (epub)

Contents

Tables

Abbreviations

ABM	*Ansar Bayt al-Maqdis* (Supporters of Jerusalem)
AFELs	Armored front-end loaders
AFRICOM	US Africa Command
ANSAs	Armed non-state actors
AOC	Anbar Operations Command
AQ	al-Qaida
AQI	al-Qaida in Iraq
ASAs	Armed state actors
ATGMs	Anti-tank guided missiles
BM	*al-Bunyan al-Marsus*
CJTF	Combined Joint Task Force
CQAs	Close-quarter assassinations
CSF	Central Security Forces
DBIEDs	Drone-borne IEDs
DIA	American Defense Intelligence Authority
DMSC	Consultative Council of Mujahidin and Revolutionaries of Derna and Its Environs
FSA	Free Syrian Army
GF	Guerrilla formations
GIA	*Groupe Islamique Armé* (Armed Islamic Group)
GNA	Government of National Accord
HBIEDs	House-borne improvised explosive devices
HLT	High lethality tactics
iALLTR	Intelligence; absorb/recruit; loot; lead; transfer (*modus operandi*)
IDF	Israeli Defence Forces
IED	Improvised explosive device
IS	Islamic State (organisation)
ISI	Islamic State in Iraq
ISIS	Islamic State in Iraq and Sham (Levant) (*Daesh* in the Arabic acronym)

ISR	Intelligence, surveillance and reconnaissance
ISWD	Islamic State Ways of Warfare Database
IYSC	Islamic Youth Shura Council
JCP	Justice and Construction Party
JN	Jabhat al-Nusra
JTAC	Joint terminal attack controller
KNC	Kurdish National Council
LAG	Loss of Accuracy Gradient
LER	Loss Exchange Ratio
LIFG	Libyan Islamic Fighting Group
LNA	Libyan National Army
LSG	Loss of Strength Gradient
MANPADS	Man-portable air-defence systems
MFO	Multinational Force of Observers
NFC	National Forces Coalition
NTC	National Transitional Council
OIR	Operation Inherent Resolve
PKK	Kurdistan's Workers Party
PMUs	Popular Mobilisation Units
PYD	Democratic Union Party
RIP	Repression, informants/intelligence and propaganda
SCAF	Supreme Council of Armed Forces
SCCLC	Soften-creep; coalition-build; liquidate–consolidate (*modus operandi*)
SDF	Syria's Democratic Forces
SGFs	Suicide guerrilla formations (*inghimassiyun* or plungers)
S/MBIEDS	Suicide/Motorcycle-borne improvised explosive device
SOSRA	Suppress, obscure, secure, reduce and assault (breaching-sequence)
SP	Sinai Province
SSI	State Security Investigations
S/VBIEDs	Suicide/Vehicle-borne improvised explosive device
TJI	*al-Tawhid wa al-Jihad* in Iraq (Monotheism and Struggle in Iraq)
TJS	*al-Tawhid wa al-Jihad* in Sinai (Monotheism and Struggle in Sinai)
UCAVs	Unmanned combat aerial vehicle
UNSC	United Nations Security Council
YPG	People's Protection Units
YPJ	Women's Protection Units

Acknowledgements

Researching the combat effectiveness, tactical innovations and strategic shifts of a group of violent hyper-extremist organisations can be difficult, dangerous and draining. To all those who helped me during this journey, I owe my deepest and most heartfelt gratitude.

I am grateful to Dr Azmi Bishara, the General Director of the Arab Centre for Research and Policy Studies (ACRPS), for his support of this research project. Thanks are also due to Dr Mohammad Almasri, the Executive Director of ACRPS, Professor Yasir Suleiman Malley and Professor Abdelwahab El-Affendi, the former and the current Presidents of the Doha Institute (DI).

I am also grateful to my research team in the Arab Centre's Strategic Studies Unit (SSU): my assistant Sofia Hnezla for the bibliographic assistance and related tasks; Dr Muhanad Seloom, Dr Massaab al-Aloosy, Abdou Moussa and Mahmoud al-Hosain for their work on the Islamic State Ways of Warfare Database (ISWD); and Abby Lewis and her team for editorial assistance.

I owe thanks to the DI's Research and Grants team: Raed Habayeb, Miriam Shaath and Manar Arafeh for their support throughout this project. ISWD and many of the data-gathering activities, including some of the fieldtrips and interviews, were made possible by the Doha Institute for Graduate Studies Major Research Fund grant (MRF 03-01). The findings achieved herein are solely my responsibility.

The research for this book initially developed during my tenure at the University of Exeter. Thanks are due to Professors Gareth Stansfield, Jonathan Githens-Mazer and William Gallois.

I am also thankful to my copy-editor, George Macbeth, and to Kirsty Woods, Bekah Dey, Eddie Clark and Adela Rauchova of

Edinburgh University Press and all their colleagues for their support throughout the writing and the editing processes. I would also like to thank two anonymous peer-reviewers for their insightful comments and critical feedback.

I am indebted to many people in Iraq, Syria, Libya, Egypt, the greater Middle East, Europe and North America for their comments, feedback and assistance with the interviews I have conducted. To name a few: on Iraq, thanks are due to Dr Haider Said, Dr Yayha al-Kubaisi, three former Iraqi army officers and others; on Syria, I am grateful to Hamzeh Almoustafa, Osama Abu Zeid, Mohammad Sarmini, Moham-mad Okda, Saad al-Shari', Matar Ismail, Nawaf Khalil and others; on Libya, thanks are due to Mohamed A. al-Darrat, Major-General Youssef al-Mangush, Anas El-Gomati, Noman Benotman, Dr Ahmed Hussein and others; on Egypt, I am grateful to Abdou Moussa, Hani Fathi and others. Thanks are also due to Lieutenant-General Terry Wolff, Anna Adam Ramadan, Camille al-Tawil, Hugo Kaaman, and Dr Aaron Zelin for their help and support; the *Jihadology* website in particular assisted in accessing many of ISI/ISIS/IS historical publications, audio- and video-releases. All the aforementioned and many others generously helped or shared their knowledge on the subject matter with me.

This book is about combat tactics and ways of warfare. So, these acknowledgements would not be complete without reference to the people who lived through the political violence and have suf-fered under the repression of dictatorships and the terror of ISIS/IS. Though I do not mention many of them by name, I am very grateful to all those who shared and trusted me with their rich experiences, extraordinary insights, inconceivable pains and modest hopes – some under conditions of almost unimaginable hardship. I hope they will accept my gratitude. And of course, any errors are mine entirely.

My final gratitude extends to my family. To my late father, my mother and my uncle, who never lost faith in me during the journey. To my uncle in London, whose insight and experiences have always enriched my thoughts. And to my sister and brother, for their love and encouragement.

Foreword

Omar Ashour examines a critical question thus far little understood in academia and even by senior military officers, which is why and how ISIS has been successful even though outnumbered and outgunned by substantially stronger state militaries, to include the huge (77 countries and 5 intergovernmental organisations) Global Coalition to Defeat ISIS. Despite a burgeoning literature on the wars in Iraq and Syria since 2011, no one has been able to offer a convincing and comprehensive explanation for why and how ISIS has been successful until now. Hence, this book solidly fills a crucial lacuna in military, war, strategic and security studies.

Beginning with the academic study of insurgency dating back to Lawrence, Mao, Templar, Lansdale, Guevara, Galula, Hoffman and more recent scholars such as Arreguín-Toft, Lyall and Wilson, Kilcullen, and Nagl; this book explores traditional factors associated with insurgency success, such as the support of an external power, popular support from the disaffected population, sanctuary, geography or topography, regime type, or other factors, which might, individually or in combination, be explanatory of ISIS success. Most of those factors are found not to be especially significant, so the author focuses his attention on the military strategies and tactics employed by ISIS and central to its successes. The author's research was based on fieldwork, interviews with soldiers and fighters who fought ISIS and extensive analysis of ISIS publications and multimedia productions. Operating mainly at the meso-level with a focus on both ISIS strategies and tactics, the analysis focuses on 17 urban battles across 10 fronts in four countries, finding that a flexible shifting across three major types of strategies, 15 categories of tactics, and multiple operational approaches best explained ISIS success. Its agility and adaptability are why ISIS was able to succeed despite its disadvantages, and this book explains precisely how that agility and adaptability works. Overall, the findings are groundbreaking.

Ultimately, readers will increase their understanding of how and why ISIS fights and how it has been so resilient and successful, but even more than that, the book provides insights into how armed nonstate actors will fight and win battles in an age of hybrid warfare. That makes this book of great interest to scholars and soldiers alike, as both need the insights it provides.

Larry P. Goodson, US Army War College

1

Is It Mainly Tactics?

If a jayvee team puts on Lakers uniforms that doesn't make them
Kobe Bryant.

> Barack Obama on ISIS' performance in
> the battle of Fallujah, January 2014[1]

Either it expands or perishes.

> Azmi Bishara on ISIS, 2018[2]

Look at what he had to fight against and look at what he did when
he fought against it.

> Mike Tyson on combat analysis, May 2020[3]

The military endurance of the 'Islamic State' organisation

When examining the 'Islamic State' Organisation (IS), it is clear that
its strength and power ratios pale in comparison to those of its state
and nonstate foes. Yet notwithstanding this disparity, and despite
the vast military force that has been deployed against it over several years by various militaries including that of the United States,
it became the dominant Jihadist organisation and once controlled
territories occupied by an estimated 8 million people in Iraq and
Syria. The organisation expanded its reach from the Syrian governorate of Aleppo to the Iraqi governorate of Diyala, a more than
600-mile-wide area, that included parts of the governorates of
Anbar, Nineveh, Kirkuk and Salahuddin in Iraq, as well Raqqa, and
parts of Hasaka, Deir Ezzor, Aleppo, Homs and Rif Dimashq (Rural
Damascus) in Syria. The organisation also occupied the suburb of
al-Hajar al-Aswad and large parts of the Yarmouk refugee camp in
Damascus and held and denied positions less than three miles away
from Souq Al-Hamidiyah in Central Damascus. IS also controlled

and/or denied territories outside of Iraq and Syria, including small parts of Libya, Egypt (northeastern Sinai) and elsewhere.

In 2014, the CIA estimated that IS' manpower numbered between 20,000 and 31,000 fighters.[4] At that time, these estimates meant that there were about ten Iraqi soldiers to every IS fighter, to say nothing of supporting and/or allied forces including Peshmerga Units, *al-Hashd al-Sha'abi* (Popular Mobilisation Units or PMUs) militias, *al-Hasha al-'Asha'iri* (Tribal Mobilisation Units) militias and the international coalition forces. Moreover, the disequilibrium in the total number of fighters was also evident in engagements between IS and its enemies. For example, on 10 June 2014, the city of Mosul, in and around which tens of thousands of Iraqi army soldiers and policemen were based, fell to an ISIS attacking force that ranged between 800 and 1,100 fighters.[5] ISIS/IS eventually lost the city of Mosul after more than nine months of fighting against American- and Iraqi-led forces that massively outnumbered and outgunned the organisation.[6]

Mosul was not the only such case. IS also managed to pull off military upsets in other Iraqi towns, such as Fallujah and Ramadi. In Syria, IS and its predecessor, the Islamic State in Iraq and al-Sham[7] (ISIS), fought against almost every other armed actor, state or nonstate, in the war-torn country. In Raqqa, ISIS was able to initially control and then fully capture this provincial capital by January 2014. ISIS was outnumbered and outgunned by a de-centralised, motley coalition of armed Syrian opposition, including units from the Free Syrian Army (FSA), *Ahrar al-Sham* (Freemen of the Levant) and *Jabhat al-Nusra* (The Support Front – JN). Despite that, ISIS units won a decisive victory in the city. It was followed by military advancements in other governorates against forces loyal to the Assad regime as well as all types of Syrian armed opposition between 2014 and 2017.[8] In Libya, IS units managed to take over the city of Sirte by March 2015, initially with fewer than 600 combatants.[9] Between March and May 2015, IS repelled a counteroffensive to liberate the city launched by the 166th Battalion loyal to the Libyan Chief of Staff and the Tripoli government. IS was also outnumbered and outgunned back then. The organisation managed to brutally suppress a local rebellion in Sirte between September and October 2015. Moreover, IS in Libya controlled several neighbourhoods in the cities of Derna and Benghazi, before losing all of the aforementioned territories by the summer of 2017.[10] In the Sinai of Egypt, differential manpower ratios were even more dramatic. In some cases, almost 100 soldiers existed for every IS insurgent in the

Sinai.[11] These are only manpower estimated ratios. None of the other combat-relevant factors – such as firepower ratios, quality of weaponry, quantity of ammunitions, air-support, intelligence-support and strategic regional and international military aid – favoured ISIS/IS.

In terms of combat effectiveness – and ignoring for now the hyper-extremist ideology, ultra-sectarian narrative, sustained brutality and the general evilness of the organisation[12] – IS has managed to accomplish sustained military upsets with relatively limited numbers and resources in the face of far superior state and nonstate armed actors. Moreover, the organisation has shown higher endurance compared to other armed nonstate actors (ANSAs) in the region, whether Jihadist or non-Jihadist. In December 2015, the Pentagon announced that it had carried out 8,600 airstrikes and dropped more than 28,000 bombs on the positions of IS in Iraq and Syria. This constituted an average of 17 airstrikes and 60 bombs dropped every day for nearly 18 months since the start of the campaign. As outlined in the following chapters, these figures pale when compared to those from 2016 and afterwards. By February 2017, IS losses were estimated to total over 45,000 members or about 50-deaths per day.[13] The Loss Exchange Ratio (LER) in some of the 2017 battles exceeded 20:1 against IS. This is compared to a 3:1 LER in the first Indochina war in favour of the French forces and their allies, and a 10:1 LER in the second Indochina war in favour of the US forces and their allies.[14]

Yet despite such losses in Mosul, Ramadi, Tikrit, Fallujah, Tal Afar, Raqqa, Deir Ezzor, Manbij, al-Bab, Jarablus, Derna, Benghazi, Sirte, Sheikh Zuweid and over 120 other towns and villages ranging from the Southern Philippines (City of Marawi) to Western Libya (City of Sabratha), the combat effectiveness of the organisation merits analysis and understanding. Up to the time of writing of this book, IS was not destroyed. Rather, it managed to endure, shift, innovate and expand elsewhere, including in Mozambique and the Democratic Republic of Congo (DRC). The IS military expansion in many of the abovementioned cities and towns defied any conventional military forecasts, taking the balances of power into account. By one measure in particular, IS proved itself far more resilient than, say, either the Taliban regime in Afghanistan or the Baath-led government in Iraq. The Taliban lost control of its *de facto* capital of Kandahar, after about 60 days of the American-led airstrikes and the Afghan opposition ground-attacks.[15] Saddam Hussein's regime lost control of its capital, Baghdad, less than 30 days after the Anglo-American invasion.[16] In the case of IS, the organisation lasted over 1,065 days[17] in both of its 'capitals', Mosul in Iraq (up to July 2017) and Raqqa in Syria (up to October 2017), under airstrikes and ground attacks by an international coalition of both state and nonstate armed actors that started on 15 June 2014 (as Operation Inherent Resolve – OIR) and 10 October 2014 (as the Combined Joint Task Force or CJTF-OIR anti-IS international coalition).

How can the endurance and the earlier expansion of such an organisation be militarily explained? How did ISIS/IS fight much stronger international, regional and local forces between 2013 and 2020? How can we account for its combat effectiveness? The following chapters of this book engage with these questions with a focus on Iraq, Syria, Libya and Egypt.

Why the weak beat or survive the strong in combat: a critical review

In many ways, IS poses a challenge to the existing literature on how and why weaker armed nonstate actors (ANSAs) beat or survive stronger armed state actors (ASAs). IS is not a unique case (or more accurately 'group of cases') in posing that challenge, yet the reasons for its resilience are; as this chapter and the following ones explain.

Since the last quarter of the twentieth century, there has been a steady rise in insurgent military effectiveness and capacities. Hence, the likelihood of insurgent victories has significantly increased. Mack (1975), Arreguín-Toft (2001), Lyall and Wilson (2009), Connable and Libicki (2010), Jones and Johnston (2013), Kilcullen (2013), Nagl (2014), Schutte (2014), Jones (2017) and other scholars have shown a rise in the victories of insurgents over stronger incumbents or in the inability of incumbents to defeat much weaker insurgents.[18] This represents a change in historical patterns. In a study of 197 asymmetric conflicts, Arreguín-Toft (2001) argued that 55 per cent of militarily weaker actors[19] were victorious between 1950 and 1998, as opposed to only 11.8 per cent between 1800 and 1845 and 34 per cent between 1900 and 1945.[20] Lyall and Wilson (2009) showed that in 286 insurgencies between 1800 and 2005, the incumbents were only victorious in 25 per cent of them occurring between 1976 and 2005.[21] This is compared to 90 per cent incumbent victories between 1826 and 1850.[22] Connable and Libicki (2010) reproduced a similar finding, after studying 89 insurgencies. In 28 cases (31 per cent), the incumbents' forces won; in 26 cases (29 per cent), the insurgents' forces won. The outcome was mixed in 19 cases (21 per cent).[23] Jones (2017) showed that insurgents were either victorious or fought incumbents to a draw in almost two thirds (64 per cent) of 181 armed conflicts since 1945.[24]

Overall, regardless of the dataset employed and the timeframe selected, the findings have been consistent. ANSAs have been altering a historical trend. It can no longer be taken for granted that the state monopolises the means of violence. Thus, it cannot be assumed that ASAs are universally more capable of defeating ANSAs on the battlefield. On the surface, it might seem that IS is simply one more organisation which fits this developing trend. Yet, the explanations of their military upsets between 2013 and 2017, as well as their resilience after facing foes with much superior man-, fire- and airpower is incomparable to most of the cases listed in the datasets mentioned above.

Still, the literature on ANSAs provides a wide range of explanations regarding military upsets, most notably those centred on population and local support, geography, external support, regime type, tactics, strategies and strategic objectives, organisational structures, and even peculiar features. The first three explanations will be referred to as the 'traditional variables' since they constitute classic explanations of numerous insurgent victories.

Popular support and regime-type

Mao ([1938] 1967) highlighted the centrality of population loyalty for a successful insurgency by stating that an insurgent 'must move amongst the people as a fish swims in the sea'.[25] The U.S. Army/ Marine Corps Counterinsurgency Field Manual concludes that insurgencies represent a 'contest for the loyalty' of a mostly uncommitted general public that could side with either the incumbents or the insurgents, and that success requires persuading this uncommitted public to side with the incumbents by 'winning their hearts and minds'.[26] Lansdale (1964), Thompson (1966), Kalyvas (2006), Kalyvas and Kocher (2007), Condra and Shapiro (2012) show that regime type matters and affects the behaviour of the population. McColl (1969) concluded that armed 'revolutions occur when there is no other means of open or legal political opposition'.[27] Based on the Vietnam experience, Lansdale (1964) understood that without popular support there was 'no political base for supporting the [anti-insurgent] fight'. People had to be convinced that their lives could be improved through social action and political reform, which requires a responsive, non-corrupt government and well-behaved armed forces.[28]

On the other end, the brutality of the incumbents against local population affects their loyalty, and therefore such brutality helps the insurgents in terms of recruitment, resources and legitimacy. General Stanley McChrystal, the former commander of the US forces in Afghanistan, refers to this effect as the 'insurgent math': for every innocent local the incumbents' forces kill, they create ten new insurgents to fight them.[29] Kilcullen (2009) earlier coined the term 'accidental guerrilla', a reference to the consequences of indiscriminate repression leading elements of the local population to be drawn into fighting the incumbents, without being prior enemies of them.[30] Walter (2014) demonstrates that the scale of corruption and repression in political institutions are not only directly related to civil war initiation, but are the primary determinants of whether or not countries get caught in locally supported, enduring insurgencies.[31] Sir Gerald Templer, who led the containment of the ethno-leftist insurgency known as the 'Malayan Emergency', summarised the salience of popular support: 'the shooting side of the business is only 25 per cent of the trouble . . . and the other 75 per cent lies in getting the people of this country behind us'. For Templer, the answer is not about 'pouring more troops into the jungle but in the hearts and minds of people'.[32] Thomas Edward Lawrence of Arabia (1929) observed that intense active support is not essential for a successful insurgency. All that is needed is 'two per cent of the population active in a striking force, and 98 per cent passively sympathetic' or, at least, neutral.[33] As elaborated upon in the following chapters, the perception of IS as a 'lesser evil' compared to the incumbents or to other insurgent forces was crucial in creating what Lawrence had already concluded in the 1920s.

Finally, when it comes to regime types and population support, a growing body of literature on armed rebellions, argued that 'strong' governments can deter the population from supporting an insurgency, thus decreasing the probability of its initiation.[34] 'Strength', however, is not clearly defined. Some of the selected case-studies in this body of literature imply that 'brutality', rather than strong professional institutions and good governance in a consolidated or a mature democracy, is a proxy for 'strength'.[35] Brutality, however, does not explain too many cases. After all, the regimes of Saddam

Hussein in Iraq, Bashar al-Assad in Syria, Mu'ammar al-Qaddafi in Libya, Abdel Fattah al-Sisi in Egypt, Anastasio Somoza in Nicaragua, Fulgencio Batista in Cuba, Mohammad Reza Pahlavi in Iran and many others were by no means weak regimes in terms of their financial capacities and their security and military institutions did not lack brutality. Their supposed 'strength' either directly contributed to the initiation of insurgencies and/or to the victory of the insurgents fighting against them.

Keefer (2008) and Getmansky (2012) consistently found that democracies are less than half as likely to experience insurgency onsets compared to non-democracies.[36] When insurgencies begin, Engelhardt (1992) and Wells (2016) empirically show that democracies have only a slightly more successful record in counterinsurgencies than non-democracies (37.5 per cent versus 33 per cent in Engelhardt's sample of 25 insurgencies and 55 per cent versus 53 per cent in Well's sample of 286 insurgencies).[37] Other studies show that when incumbents in a democracy end up fighting insurgents, their performance in counterinsurgencies and the outcome of the conflict varies depending on other factors, including strategies and tactics employed by the belligerents, population-support, geography, external support, and not necessarily regime type.[38] However, the type of the regime influences the incumbents' and the insurgents' initial decision to become involved in the armed conflict.[39]

Geography

Geography-centric explanations can also be found in the literature. As early as the 1920s, Thomas Edward Lawrence of Arabia observed that guerrilla warfare would most likely be effective when conducted from rough and inaccessible terrain in a country's interior.[40] Mao ([1938] 1967) concurred.[41] He also argued that guerrilla warfare is most feasible when employed in large countries where the incumbents' forces tend to overstretch their lines of supply. Macaulay (1978) and Guevara (1961) showed how small numbers of armed revolutionaries in Cuba manipulated the topography to outmanoeuvre much stronger forces and gradually moved from the eastern provinces of the island towards the capital in the West.[42] Guevara (1961) in particular emphasised the importance of escaping the incumbents' reach through utilising the rough terrain in parts of Eastern Cuba:[43] 'fighting on favourable ground and particularly in the mountains presents many advantages' for the guerrillas.[44] Galula (1964) was more deterministic when it came to geographical explanations. In his seminal work *Counterinsurgency Warfare*, he stresses that "the role of geography . . . may be overriding in a revolutionary war. If the insurgent, with his initial weakness, cannot get any help from geography, he may well be condemned to failure before he starts'.[45] McColl (1968) outlined a profile of geographical zones hospitable to successful insurgencies. It has seven features, including a history of political instability, access to important political or military centres such as provincial capitals or regional cities, economic self-sufficiency, a rugged terrain and a complex international boundary.[46]

Boulding (1962) introduced the concept of the 'Loss of Strength Gradient' (LSG) to geographical explanations.[47] Briefly, it means that the further the fight is from the centre, and the deeper it is into the periphery, the more likely for the incumbents' forces to lose strength. Schutte (2014) builds on and modifies the concept to argue that it is accuracy, not necessarily strength, which gets lost as a function of distance.[48] He introduces the 'Loss of Accuracy Gradient' (LAG), positing that incumbents' long-range attacks are more indiscriminate and less accurate (in killing insurgents) than short-range ones. Hence, civilian alienation becomes a function of distance, as a result of inaccuracy and indiscriminate killings. Fearon and Laitin (2003) stressed that rough terrain is one of four critical variables supportive of an insurgency.[49] Kalyvas (2006) does not provide a geography-based model for insurgents' military effectiveness, but he argues that military victory is endogenous to geography.[50] Overall, geography is one of the most recurrent explanations of insurgents' successes or survival in the literature written by both scholars, practitioners and insurgent leaders. Nonetheless, as outlined below and in the following chapters, it is not the main explanation for the endurance and earlier expansion of IS Provinces, given the geographical natures of the areas in which they have operated, denied or occupied.

State-sponsorship

Other scholars have highlighted foreign state sponsorship as a critical variable in insurgents' victory. In their study of 89 insurgencies, Connable and Libicki (2010) argued that insurgencies that 'benefitted from state sponsorship statistically won a 2:1 ratio out of decided cases [victory is clear for one side]'. Once sponsorship was wholly withdrawn, the victory ratio for the insurgent side fell to 1:4.[51] However, this is relevant only to clear-cut victories, not to mixed outcomes or draws (when incumbents and insurgents reach a compromise or a settlement). It is also irrelevant to enduring insurgencies. Moreover, in their study of 286 insurgencies between 1800 and 2005, Lyall and Wilson (2009) defined external support for an insurgent as an international patron(s) that provided material support, a rear base across international borders to organise and train fighters and evade the incumbents' countermeasures, or both. Their study shows that 70 per cent of insurgent organisations that received external support either won or fought their way to a compromise. Groups without any external support won or negotiated a settlement in only 28 per cent of the sample.

Jones (2017) qualifies the argument by showing how 'great power'[52] support is critical for insurgent victory. In the 286 insurgencies he examines, the insurgents were victorious in 52 per cent of this sample when they received great power support. Insurgents receiving support from a bordering state and a diaspora represented only 42 per cent and 38 per cent of insurgent victories in the sample respectively.[53] Other scholars focused on the type of support and how rebels attract it.[54] Combat assistance (lethal materiel, combat support, combat training) from state actors is usually critical for the victory of weaker insurgents against foreign-supported incumbents. As shown in the cases of Afghanistan, Nicaragua and Syria, particular types of lethal materiel, such as anti-tank guided missiles (ATGMs) and man-portable air-defence systems (MANPADS), were extremely useful when insurgents faced better-equipped incumbents' forces supported by foreign powers.

Other types of external support critical for the victory or the endurance of an insurgency include sanctuary, financial, logistical and/or intelligence support.[55] The importance of sanctuary in particular has been debated among scholars of insurgency studies. Byman (2001) argues that sanctuary provided by an external state is a critical factor for insurgents' victory.[56] The literature has also discussed financial and non-lethal support by external actors. Financial support was correlated with the endurance of insurgencies and with prolonging an armed conflict.[57] It was, however, less correlated with the victory of insurgents compared to other variables. Non-lethal support, including propaganda dissemination, access to media and narrative broadcasting, had a limited impact when it came to the outcome of insurgencies as well. Intelligence support and breakthroughs were exceptions. Several cases of successful insurgencies showed how crucial intelligence/ information support by foreign state-actors could be. Identifiable cases include those of the Nicaraguan Sandinistas and the Cuban *Dirección General de Inteligencia*; the Algerian *Front de Libération Nationale* (FLN) and the Egyptian General Intelligence Directorate; and Hizbullah and the Iranian Revolutionary Guards' intelligence agencies and al-Quds Force.[58] Finally, Freedman (2015) warns not to conflate local support for an insurgency cloaked in ideological rhetoric as a response to local conditions with foreign support resulting in initiating an insurgency.[59] Such a conflation can lead to misguided policy implications and counterinsurgency strategies, which by themselves can contribute to a victory for the insurgents or the protraction of the conflict.[60]

State-sponsorship, however, neither explains the expansion nor the endurance of ISIS/IS. Intermittent coordination between international and regional powers, regional rivalries, and conflicting aims and interests of different state- and state-sponsored parties have had tactical and operational ramifications, which IS have capitalised upon sometimes. But these factors are different from direct state-sponsorship. Kalyvas (2010) highlights the gaps in the literature when it comes to explaining how 'radical Islamist' organisations fight and endure, despite a vast reserve of explanations for how the end of the Cold War has limited external support and affected Marxist and ethno-nationalist insurgencies and their 'technologies of rebellion' (fighting methods).[61] What is more puzzling in the case of IS (and its predecessors ISI/ISIS) is the fact that 'great-power' support was given to the ASAs and ANSAs combating it since its inception. The Global Coalition against *Daesh*[62] (ISIS) is composed of 77 states and five International Governmental Organisations (IGOs).[63] It does not even include states such as Iran, Syria and Russia which have also engaged the organisation. The coalition also excludes ANSAs and coalition of ANSAs which intensely fought IS such as Syria's Democratic Forces (SDF), Lebanese Hizbullah, the Iraqi Popular Mobilisation Units, various state-sponsored Libyan anti-IS ANSAs, and various Sinai-based tribal militias, among others. Unlike al-Qaida, IS did not have any state-sponsorship.

Tactics and strategies

Scholars have also explained why weaker insurgents win or survive in terms of military tactics and strategies. Three major categories exist in tactical and strategic explanations: incumbents' ineffectiveness, insurgents' ingenuity and favourable strategic interactions (for the weaker side). Tactically, Lyall and Wilson (2009) offer an explanation through incumbents' ineffectiveness. They argue that modern combat machinery has undermined the incumbents' ability to win over the civilian population, form ties with the locals and gather valuable human intelligence.[64] Hence, the tactics used undermine the overall strategic goals and effectiveness, especially if the insurgents capitalise on popular support and gather counterintelligence. Hoffman (2006), Jones and Johnston (2013), Kilcullen (2013) and Seig (2014) focus more on the ingenuity of the insurgents and their newfound abilities to break states' traditional monopoly on the means of violence. Since the end of the Second World War, insurgents were gradually able to access new technologies in weaponry, communications, intelligence, transportation, infrastructure and organisation.[65] This allowed them to enhance their military tactics to levels historically reserved for state-affiliated armed actors. And that result significantly offset the likelihood of a defeat inflicted on them by the incumbents' forces.

Others specifically focused on the quality of insurgents' tactics, as an explanation for strategic upsets. Sustained ambushes, raids, subversions, sabotage operations, assassinations and various type of bombings can – at the very least – exhaust the incumbent's political, as opposed to military capability to fight.[66] General Alberto Bayo of Cuba and Spain – a mentor of both Fidel Castro and Che Guevara – stressed the importance of surprising ambushes in particular:

> every good guerrilla must rely on surprise, the skirmish, the ambush and always attack when the enemy is confident and does not expect attack. When the enemy begins to counterattack, we must disappear from sight and withdraw to a safer place.[67]

Engelhardt (1992) shows how reliance on harassment and surprise, stealth and raiding can lead to territorial control in peripheral areas, either by destroying the will of the incumbents' forces to fight or by using these tactics to transform into a conventional military strategy, and thus destroying the capacity of the incumbents to fight.[68] Mao (1937 [1967]), blends the two approaches. He stresses that the guerrilla tactics should aim to grind down the incumbents' will, before launching a major conventional attack to destroy their capacities. Jones (2017) argues that there is no clear correlation between the use of a particular tactic and achieving a military upset for the insurgents in their overall campaign. The conclusion, however, neither fits nor explains many of IS (and other insurgents') victories as demonstrated in the following chapters.[69] Other scholars have shown how high-lethality tactics (HLTs) can impact the overall outcome of the conflict. Suicide tactics in particular were focused upon, including the utilisation of belt-, boat-, car-, motorcycle-, truck- and mixed-bombs.[70] One study observed that organisations which use suicide tactics were either defeated or are still engaged in an on-ongoing conflict. None attained a victory, so far in the sample employed in the study.[71] However, the cases of the Chinese revolutionaries 'Dare to Die' Squads (between 1911 and 1949),[72] Hizbullah (between 1985 and 2000) and Hamas in 2007 (Battle for Gaza) are ones in which organisations utilised suicide-tactics and either secured an outright military victory or attained an otherwise favourable outcome. The Taliban – by 2020 – has conducted over 1,000 suicide attacks. The organisation has fought and suicide-bombed its way to an official compromise with the United States.[73] Also, IS managed to pull off military victories in specific phases of its insurgencies, as outlined in the next four chapters while heavily relying on various categories of suicide tactics. A final note on tactics should be mentioned here. The guerrilla tactics of hit-and-run are not necessarily a sign of weakness or cowardice, as some of the conventional counterinsurgency analyses in the literature imply. The use of these tactics reflects strategic calculations by the insurgents to enable them to fight and, sometimes, to defeat much stronger foes. They are no more 'cowardly' than the use of aerial bombardment in order to avoid casualties by refusing to challenge the insurgents in a ground battle. The decision to use aerial bombardment is not an indicator of cowardice; instead, it reflects strategic calculations and utilisation of the unique capacities available to the incumbents and their ASAs.

Scholars have also explained how insurgents achieve military upsets in terms of their military strategy and strategic interactions with their enemies. Arreguín-Toft (2001) produced a complex model of strategic interactions between militarily weaker actors and their stronger enemies. His study concludes that weaker forces can overcome resource paucity by employing opposing strategies (direct versus indirect) against stronger ones. A guerrilla warfare strategy (an indirect strategy) is the most suitable to employ against direct attack strategies by stronger actors including 'blitzkriegs'.[74] As a result, he showed that strong actors won 76 per cent of all same-approach strategic interactions in his dataset, while weak actors won 63 per cent of all opposite-approach interactions.[75] Another strategy-based explanation is 'strategic-shifting': when insurgent organisations alternate strategies – based on their resources – to execute a military upset against a stronger incumbent enemy, or to achieve a favourable outcome. Three types of strategies are usually employed: conventional, guerrilla and terrorism strategies. Engelhardt (1992) argues that insurgents usually win by transforming themselves from a guerrilla to a conventional strategy and defeating the enemy in the field (like Mao Tse-tung's forces in China in the late 1940s).[76] Other scholars linked the success of the insurgents to the strategies employed, but this remains an inconclusive debate in the literature.[77]

A related explanation to the strategic *objectives* of the insurgents and how they impact their odds of victory. Three insurgents' strategic objectives have been analysed in the literature: local regime-change, secession/irredentism, and anti-occupation/national-liberation objectives. Whereas anti-occupation/national-liberation objectives correlate more with insurgent victories, secession has a lower rate of success compared to other strategic objectives.[78] Finally, scholars debated how maximalist goals affect the outcome of an insurgency.[79] Although most insurgencies were decided on the battlefield, many ended up in draws or compromises.[80] Insurgents who have alternative goals (or a prioritised repertoire of goals), who have the capacity to politically manoeuvre and to compromise to lessen the odds of their defeat by allowing chances for a settlement. Mao stated that 'without a political goal, guerrilla warfare must fail, as it must, if its political objectives do not coincide with the aspirations of the people'.[81] Maximalist goals, however, still serve as a good tool for recruitment and radicalisation, boosting fighting morale, and outbidding rival insurgents by branding them as 'sell-outs'.[82]

Organisational structures and peculiar features

Finally, among the variables analysed in the literature and impacting the odds of victory or defeat are the organisational structures and peculiar features of the (state and nonstate) belligerent forces. Levels of centralisation, decentralisation, factionalism and other organisational and peculiar variables can affect the combat performance and overall military effectiveness of both the insurgents and the incumbents' forces, and thus influence the probability of victory or defeat.[83] Pollack's (2019) seminal work on Arab ASAs has a section in the last chapter before the conclusion that analyses why Hizbullah and ISIS/IS have fought 'considerably better than most Arab armies of the modern era, state or nonstate'.[84] Hizbullah can be explained by the 'traditional variables' and others in the above review. The organisation enjoyed almost four-decades of intense and sustained state-sponsorship. It had, and continues to have, popular support among large segments of the Lebanese society. Especially between 1992 and 2012, Hizbullah also had popular support among large segments of other Arab-majority societies that ebbed and flowed but consistently transcended ideological, sectarian and religious faultlines. ISIS/IS had none of that.[85]

To address the case(s) of IS, Pollack (2019) proffers a complex argument of six peculiar features and conditions: weak and incompetent adversaries, zeal, 'Darwinian selection' of IS members, unorthodox (de-centralised) hierarchies, foreign fighters and manipulation of (Arab) cultural strengths.[86] All six certainly mattered and contributed to the IS' military expansion and endurance.[87] It is accurate to conclude that IS has *mainly* faced corrupt and combat-mediocre foes by any Western military standard. However, IS was certainly much *weaker* than them as explained in the following chapters. Also, as detailed later, the aforementioned peculiarities were not so peculiar; they were shared and exhibited by other armed nonstate actors who never attained any similar successes whether in Iraq, Syria, Libya and/or Egypt. Globally, from the Tupamaros and the Guevara-led National Liberation Army in South America to the *Groupe Islamique Armé* (GIA) in Algeria and the Moro Islamic Liberation Front in the Philippines, other ANSAs exhibited some or all of these features in addition to other aforenoted supportive variables such as geography and external support. Yet, they never reached the same level of expansion or endurance compared to IS. The GIA offers a good comparative contrast. An earlier ideological formation akin to IS, the GIA had zeal-on-steroids,[88] de-centralised smaller formations, experienced foreign fighters,[89] cultural strengths and nonconformities, in addition to a supportive geography and, at the very onset of its insurgency, some local and external support.[90] It also had poor combat performance, low-to-no military effectiveness (depending on which combat unit or guerrilla formation) and almost no military feats comparable to IS'.[91]

Situating the IS case-studies in the literature

The explanations reviewed above have made important contributions to the expansion of knowledge about how weaker insurgents can defeat stronger incumbents or fight them to a stalemate and/or a compromise. Several elements of these explanations are applicable to the various cases of IS Provinces in Iraq, Syria, Libya, Egypt and elsewhere at different stages and points in time, most notably the LSG, the LAG and both insurgents' and incumbents' tactics and strategies. Still, the endurance and expansion of IS Provinces (and other ANSAs) deviates from the above review and poses a challenge to some of the explanations in the wider literature. On a macro-level, the socio-political environment in the Arab-majority world has some particularities. A combination of arms and religion or arms and chauvinistic ethnonationalism has proved to be the most effective path to gain and remain in political power in most of the Arab-majority states. Votes, constitutions, good governance and socio-economic achievements are secondary variables and in many Arab countries, they are relegated to the margins. IS can certainly endure and expand in a regional context where bullets prove again and again that they are much more effective than ballots, where extreme forms of political violence are committed by state and nonstate actors and then legitimated by religious institutions, and where the eradication of the 'other' is perceived as a more legitimate and 'patriotic' political strategy than compromises and reconciliations. On a meso-level, as elaborated upon in the next few chapters, IS and its predecessors have mastered diversifying and shifting their strategies. The constant shift from/to conventional, guerrilla and terrorism ways of warfare became a hallmark of the organisation. Tactically, IS was able to modify and upgrade tactics associated with urban terrorism (such as vehicle borne IEDs or VBIEDs) and use them in a way similar to conventional weapons, such as artillery and guided rockets. Their training camps and manuals show a strategic choice made by the organisational leadership to invest in teaching its members the importance of shifting strategies and the innovative tactics associated with each of these strategies. On a micro/individualist level, IS has been able to train and empower its fighters, in a way that many Arab armies were incapable and/or unwilling to do.[92] This has been the case especially when it comes to de-centralisation and levels of autonomy on tactical and operational levels.

Explaining the puzzle

The main two research question that this book aims to answer are how did IS fight and why did it militarily endure and expand in Iraq,

Syria, Libya and Egypt? This section elaborates more on the definitions of the terms used throughout the book, and on the hypotheses
and the methodology adopted to engage with the research questions. The cases of IS 'provinces' in Iraq, Syria, Libya and Egypt are
quite unique because their endurance and expansion defies many of
the reviewed explanations in the literature, including most of those
centred on geography and popular support of the insurgents and all
of the explanations centred on state-sponsorship of the insurgents.
The book primarily focuses on the insurgent side. It neither focuses
on the incumbents' forces nor on their counterinsurgency and counterterrorism policies and practices. It attempts to explain a particular
dimension of IS insurgencies, which is how they fought and why
they managed to beat or survive much stronger state and nonstate
actors.

Definitions and terminology

This section is composed of two parts. The first part defines some of
the combat-relevant terms used throughout the book. The second
part defines some of the ideological terms, given the background of
IS and the ideologically laden rhetoric it uses. The acronyms IS, ISIS
and ISI (Islamic State in Iraq) refer to the Islamic State organisation
and two of its earlier predecessors. Throughout the book, the usage
of these acronyms depends on the timeframe being discussed. ISI
existed between October 2006 and April 2013, before it becomes
ISIS. ISIS only lasted between April 2013 and June 2014, before it
becomes IS.

In terms of combat-relevant definitions, 'conventional warfare'
here refers to direct military confrontations across defined frontlines between armed and, sometimes, uniformed units. It typi-
cally involves major combat operations that overtly seize control
of territory, inhabitants and resources with the strategic objective
of destroying or subduing the enemy. 'Guerrilla warfare' refers to
a form of unconventional warfare used by smaller formations of
combatants to avoid direct military confrontation with their foes.
These formations heavily rely on indirect and surprise tactics such
as ambushes, sabotage, raids, petty warfare and hit-and-run to fight
a larger and less-mobile enemy. The strategic aim of guerrilla warfare is attrition or the destruction of the will of the enemy, not necessarily its capacity, to fight. 'Terrorism' here refers to a repertoire

of armed tactics by which civilians and/or non-combatants are violently targeted, discriminately or indiscriminately, for the strategic aim of intimidation or pressuring their rulers (state or nonstate) to accede to political and/or ideological demands. In many ways, terrorism mirrors coercive strategic bombardment whereby civilians are targeted and punished to coerce their governments or rulers.[93] 'Insurgency' is an armed and organised rebellion against an existing political status-quo, with the aim of toppling it. Multiple types and tactics of conventional, guerrilla and terrorism warfare can be used during an insurgency.

'Tactics' refer to the art and science of utilisation and organisation of force on the battlefield during engagements with the opposing side(s) or in close proximity to it/them; with the aim of translating combat skills and resources to a position of advantage or an outright battle victory. All categories of tactics employed during the battles and the campaigns discussed in this book are based on organised formations (such as suicide guerrilla formations, SFGs or *inghimassiyun*), weapon systems (such as MANPADS), or enablers (such as tunnels). All tactics discussed are based on techniques and procedures and combine two or more of the following four features: firepower/explosive-power, mobility-positioning, protection-stealth and shock. The operational level connects combat tactics to the overall war strategy.[94] It encompasses a sequence of tactical actions and *modi operandi* with a unifying objective, in a series of interrelated operations/battles/combat theatres with the aim of achieving an advantage in a campaign and getting closer to a strategic victory. Tactical and operational levels directly impact combat effectiveness. The latter reflects the quality of the performance of a fighting force on the battlefield, usually based on its skill, will and other factors. 'Combat effectiveness' is one aspect of the overall military effectiveness of an armed organisation. 'Military effectiveness' is a wider concept reflecting the quality of the military performance on a strategic level. It describes the capabilities to translate combat and non-combat resources into military power to achieve the strategic or grand-strategic objective(s). It should be stressed here that neither combat effectiveness nor the wider concept of military effectiveness are sole determinants of the battle- or the war-outcomes. Cleary, they are among the main factors influencing outcomes, but other factors can and will interact with them to engender battle- or war-outcomes between state and nonstate armed actor(s).[95] These factors include some of the aforenoted ones such as geography, state-sponsorship/interventions and popular support. This is in addition to simpler quantifiable and non-quantifiable factors such as the sheer numbers of soldiers/fighters, quality of weaponry, quantity of ammunitions, morale/'moral force', suitable organisation, intelligence capacities, maintenance and logistics, topography, climate and even flukes. Hence, highly combat-effective forces can still be defeated by larger forces or by peculiar conditions and thus fail to achieve their tactical, operational and/or strategic objectives. The *Wehrmacht* and the Japanese Imperial Armed Forces – two superbly combat-effective armies – are classic modern examples, among others. This brings us to the definition of 'strategy': the long-term organisation, distribution, planning and application of all combat and non-combat resources to serve and achieve the ultimate objective(s) of the entity(ies) discussed, whether state or nonstate.[96]

Given the ideological variations within Islamism and how IS relates to them, it is useful to define some of the Islamist ideologies that will be referred to throughout the book. 'Islamism' is a set of often conflicting political ideologies that base and justify their principles, behaviours, strategies, organisational structures and objectives on a certain understanding of Islamic texts or on a certain past interpretation of Islamic texts. 'Jihadism' is a modern Islamist ideology which believes that armed confrontations with political rivals are the only theologically legitimate and instrumentally efficient method for socio-political change.[97] Depending on the type of Jihadists, combat (*qital;* which is equivalent to *jihad* in this case) can be *the* legitimate means for change, or an end in and of itself. Jihadism, as a modern Islamist ideology, should not be confused with 'jihad', an Islamic concept/duty that refers to various types of non-violent striving (such as struggling against 'sinful' desires, 'Satan', or disease) and violent struggle (such fighting against 'infidels' or invaders). Jihadism is a subcategory of 'Armed Islamism', a larger category in which Islamists can take up arms for various reasons, including the belief that arms can be *a* mean for socio-political change (including local regime change, secession/ irredentism and fighting an occupying force) or a mere defensive tactic for socio-political survival.

'Salafism' is another subset of Islamism, with at least six main ideological trends: Scholarly Salafism (*al-Salafiyya al-'ilmiyya*), Electoral Salafism, Authoritarian Salafism (such as *al-Madakhaliyya/ al-Jamiyya*), Civil Resistance Salafism (*al-Salafiyya al-Thawriyya*), Combat Salafism (*al-Salafiyya al-Jihadiyya* or Salafi-Jihadism) and

Wahhabism (al-Wahhabiyya). Ideological commitment to Salafism alone – without the six aforementioned qualifiers – does not predict political behaviour. The six Salafi ideological trends have a few commonalities, though. The first is the common belief that the first three Muslim generations (seventh to eight century ad) had the ultimate understanding of Islam and therefore their behaviours and practices should be followed as it was done in their times. According to Salafi puritanical beliefs, any religious 'innovation' that deviates from these practices is sinful and can lead to polytheism and infidelity. A second commonality is the strong emphasis on monotheism (tawhid) and what it entails in rituals, beliefs and social and political behaviours. Finally, all Salafi trends claim to uphold literal interpretations of primary Islamic sources (The Quran and collections of Hadith). However, this is a highly dubious claim.

Salafi ideologies differ on the theological concepts of takfir (accusations of disbelief) and khuruj (taking up arms against an established ruler/consolidated dynastic rule), as well as on political concepts such as electoral democracy (whether it is legitimate or illegitimate, a 'necessary evil' or an outright 'infidelity'). Wahhabism is a Salafi ideology that combines the traits of Authoritarian, Scholarly and Combat Salafism. It has three main distinctions. The first is its emphasis on the works of Muhammad ibn Abdul Wahhab, an eighteenth-century ultraconservative Salafi theologian from Najd whose actions and writings proffered the religious, political, social and combat legitimacy for the first, second and third Saudi States.[98] The second distinction is that Wahhabism has a direct affiliation with the Kingdom of Saudi Arabia. The ideology and its dogmatic tenants are adopted by the official religious institutions, and Wahhabi affiliates and loyalists ran several state-funded universities, educational institutions, official agencies, the judiciary and the moral police (Committee for the Propagation of Virtue and the Prevention of Vice). And thirdly, Wahhabism is influenced by the traditions and culture of the Najd region in Saudi Arabia, much more than other Salafi ideologies.[99]

Lastly, 'Takfirism' is an ideology whose basic assumption labels a whole Muslim community (globally or in a country, a city, a town or a village) as apostates, unless proven otherwise. Hence, a whole set of social sanctions and 'theologically legitimate' punishments can be applied against them, including targeting them.[100] Takfirism as an ideology differs from the concept of takfir (excommunication or

accusation of disbelief), which is not only used in other Islamist ideologies, but also used by religious institutions and governments on
a relatively limited scale, especially to stigmatise opposing individuals.[101] Takfirist organisations can be violent or non-violent. If they
are violent, the scale and intensity of violence is usually high, given
the wide selection of civilian and soft targets.[102]

On the Islamist ideological spectrum and its terminologies, IS
is primarily a Salafist-Jihadist organisation which extensively uses
the concept of *takfir* to legitimate its violence against Muslims and
non-Muslims, even compared to other Salafist-Jihadist groups. The
organisation relies heavily on Wahhabism's teachings in general and
the writings of the Muhammad Ibn Abdul Wahhab in particular.
That makes it in many ways akin to the nineteenth century *Ikhwan
Min Ta'allah* (Brothers of Whoever Obeys God) tribal militias,
which conquered parts of the Arabian Peninsula and was a main
force behind the establishment of the Saudi Kingdom in 1932.[103]
Arguably, IS stops short of being Takfirist as it does not declare non-
members as apostates unless proven otherwise.[104] It did punish and
even execute members who did that.[105]

Hypotheses: is it mainly about tactics?

The book proposes two main hypotheses. The first is that IS has
been able to militarily beat or survive stronger ASAs and ANSAs
due to its ability to effectively shift between three combat strategies: conventional warfare, guerrilla warfare and terrorism. The
second is that IS has been able to pull off military upsets against
stronger ASAs and ANSAs due to combining innovative tactics
associated with the aforementioned combat strategies in different
theatres.

The quality and quantity of IS tactical innovations often surprised
its enemies in urban, rural, mountainous and desert theatres. Once
the repertoire of innovative tactics was exhausted and repeated without new innovations, the surprise element had limited impact. This
is when the factors and the explanations reviewed above, such as
manpower, firepower, aerial bombardment, geography, population
support and others, became more decisive; and the capacities of IS
Provinces to fight back became significantly diminished. Concisely,
the two hypotheses in other words are:

H1: IS tactically and operationally endures or expands due to successful shifts between conventional, guerrilla and terrorism strategies.

H2: IS orchestrates operational military upsets against stronger enemies due to innovative combinations of tactics associated with conventional, guerrilla and terrorism ways of warfare.

These two hypotheses are based on a meso-level (organisational-level) analysis of IS. But two other levels of analysis matter in the explanation of how IS fought and why it endured and expanded against stronger enemies. However, they will not be the primary focus of this book due to the constraints of space. Earlier studies have also given these two levels well-deserved attention.[106] The first is the micro-level (individual-level) analysis. That level has to do in particular with the types and beliefs of the individual fighters upon which IS relies. Three critical categories of fighters who militarily matter were identified for the purpose of this book.[107] The first category includes former members of regular armed forces, including Iraqi, Syrian, Libyan, Egyptian, Georgian, Russian, Tajik, Europeans and others. The second category includes battle-hardened guerrillas who earlier fought in local or foreign insurgencies such as in Afghanistan or previously (pre-ISIS) in Iraq. The third category includes the persistent local insurgents, who accumulated significant experience of both combating the incumbents' forces and building logistical support networks over the last decade. Both local and foreign fighters can belong to any of the aforementioned categories.

In terms of beliefs and motivations of individuals and how they impact their military performance, three categories were also identified: the *oppressed*, the *dogmatist* and the *opportunist*. The first usually joined IS as a result of grievances held against the status-quo or another organisation. For the *oppressed*, IS can range from a lesser evil to a saviour and/or an avenger. And IS usually capitalises on the level of grievances in its armed activities as well as on the local knowledge of the aggrieved/oppressed. The *dogmatist* type is equally critical for IS: ideologically committed and willing to die for their cause. This type of fighter supplements and does not always contradict or clash with locally rooted, aggrieved insurgents. Finally, the *opportunist* type is also useful for IS. On one end, this type tends to avoid deadly operations, such as suicide bombings and suicidal guerrilla formations (*inghimassiyyun* or plungers) and is not as effective in desperate defensive operations. On the other end, the opportunist type is usually ambitious, rationally risk-friendly and in terms of tactical aggression, some of the fighters in that category proved to be quite effective in their military performance.[108]

The next significant level of analysis is the macro-level (national and regional contexts). ISI/ISIS/IS operated in socio-political environments that reward political violence in many forms (occupation, coups, repression, insurgencies, terrorism, and others) and punish moderation, compromises and general conciliatory behaviour. Hence, IS is a symptom, not a cause, of the deeply dysfunctional and violent politics in the region, especially in Iraq, Syria, Libya and Egypt. The aforementioned combination of arms and religion or arms and hyper ethnonationalism/tribalism/sectarianism in these countries has consistently proved to be the most effective means to gain and remain in political power. IS and like-minded organisations did no more than increase the levels of violence and further manipulate religious texts to legitimate the violence. IS has done so to make up for its material weaknesses, compared to established local ASAs.[109] IS operated and still operates in environments where transitions to non-violent conflict resolution mechanisms and constitutional, institutionalised politics have often failed to consolidate.[110]

The Arab-majority uprisings that started in Tunisia in December 2010 have provided scholars and practitioners with several important lessons on how changes within the macro political environment can affect the militarisation of politics and give rise to groups such as IS. The success of mainly unarmed civil resistance tactics that brought down two authoritarian regimes in Tunisia (2010/2011) and Egypt (2011) briefly undermined the rationale of armed radicals; that armed action is the most effective (and, in some ideologies, the most legitimate)[111] means for political change. But the brutal tactics of the Gaddafi and the Assad regimes in dealing with protestors have shown the limits of civil resistance. These limits were re-highlighted in Iraq in April 2013 (the crackdowns by al-Maliki's government on sit-ins) and in Egypt in July and August 2013 (during and in the aftermath of al-Sisi's military coup). This showed a different conclusion not missed by many young Arab activists: soft power and civil resistance tactics have their limits and to pursue real change, hard power is necessary. And as a result of that regional context, radicalisation, recruitment and ideological frames supportive of armed militancy are more likely to grow.[112] Hence, on a macro-level, IS and like-minded organisations were supported by that peculiar regional context.

Research design

Back to the meso-level, to understand and analyse how IS fights in the four countries, the book primarily employs qualitative, comparative research. It is based on fieldwork, which particularly focused on a sample of IS urban battles and battlefronts in Fallujah, Mosul, Ramadi, Raqqa (City and Governorate), Derna, Sirte and Sheikh Zuweid. The sample includes 17 battles across ten different battlefronts. Beyond the main seven governorates[113] in which the analysed battles have taken place (Nineveh, Anbar, Raqqa, Deir Ezzor, Derna, Sirte and North Sinai), the following chapters also include observations from other governorates.

The case-selection of this sample of battlefronts was based on either their strategic value for IS, such as the two 'capitals' in Mosul and Raqqa; and/or on the 'combat investment' and thus a reflection of IS Provinces' tactical and operational capacities, such as the three cases of Fallujah, Sirte and Sheikh Zuweid.[114] The comparative levels of 'combat investment' were commented on by the spokesperson of the Combined Joint Task Force (CJTF) of Operation Inherent Resolve (OIR) to fight IS, Colonel Ryan Dillon:

> When we compare Mosul and Raqqa, they – the extents of the defence network, the defences that have been put in by ISIS, you can tell where their, quote, unquote, 'twin capitals' have been, and where they've spent their time. We did not see this type – these types of elaborate defences established in Tal Afar by any way, shape or form. But I would – I would say that in Raqqa what defences they put in, with IEDs and these tunnel networks, are comparable to what we've seen in Mosul.[115]

In all of the selected battlefronts, IS innovated tactics, adapted its operations and alternated between conventional, guerrilla and terrorism strategies in mixed terrains (urban, suburban and rural). Overall, eight IS 'Provinces' were directly engaged in these battles. They are Fallujah, Anbar, Nineveh, Raqqa, al-Khayr (Deir Ezzor), Cyrenaica, Tripolitania and Sinai Provinces. At least another six provinces provided combat-relevant logistical assistance. These six were Salahuddin, Diyala, Latakia, Aleppo, Homs/al-Badiya and al-Furat Provinces. In these battlefronts, IS Provinces confronted over 80 armed nonstate actors; some of them were organised in large coalitions such as the PMUs in Iraq (over 39 militias), the SDF in Syria (about 18 militias), and the *'Shura* (Consultative) Council of the Mujahidin and Revolutionaries of Derna and Its Environs' (DMSC) and their allies in Libya (about five militias). IS also faced ground and air combat units from – at least – 22 armed state actors in these battlefronts.[116]

To gather some of the primary data employed in this book, the author conducted 58 semi-structured interviews with military and paramilitary commanders, militiamen, soldiers and civilians who were either involved in combatting IS or observed/eye-witnessed a specific battle as non-combatants. In total, 31 of the interviewees directly engaged ISIS/IS units in one of the four countries. The rest of the interviewees were involved in advising anti-ISIS/IS forces or observed and/or eye-witnessed urban battles with ISIS/IS as non-combatants. The author also used information from another 36 open-source interviews with former and active state officials from military, security or intelligence institutions who were involved in combatting or countering ISIS/IS. Given the sample bias, the author used some of the interviews of ISI/ISIS/IS published by their media outlets,[117] in addition to open-source interviews and social media observations of ISI/ISIS/IS sympathisers and defectors. To comply with the research ethics requirements, the author did not conduct any interviews with prisoners – despite many available opportunities – nor any person suspected to be under direct or indirect duress. Also, for safety reasons, some of the interviewees were anonymised based on their requests.

To understand the military build-up and development of ISI/ISIS/IS, the author and his research team reviewed 228 issues of the Arabic-language *al-Naba'* newsletter. This is the IS weekly newsletter that started on 17 October 2015 and was still circulating up to the time of writing of this book. Seventy-eight previous versions of *al-Naba'* newsletter were also reviewed by the author and his research team. They were published by ISI (and then ISIS and then IS) as of May 2010. These 78 issues were published on an irregular basis first, then on a monthly basis and then on a biweekly basis. ISI also published detailed annual military metrics in September 2012 (198 pages documenting 4,500 operations) and October 2013 (410 pages documenting 7,861 operations).[118] All 15 issues of the IS English-language *Dabiq* magazine and 13 issues of IS *Rumiyah* magazine were reviewed for relevant information and interviews. This is in addition to over 200 ISIS/IS and other armed organisations' audio- and video-releases and photographic reports relevant to the battles, battlefronts, armaments and tactics analysed in the following chapters. For verifications, comparisons and cross-checks, the author relied on the statements of, and media reports on, the US-led Combined Joint Task Force (CJTF), Operation Inherent Resolve (OIR), Operation Odyssey Lightening (in Libya), Operation Shader (by the United Kingdom), Operation Impact (by Canada) and *Opération Chammal* (by France). This is in addition to the media reports and interactions with local activists and institutions. Overall, throughout the following chapters, multiple data sources and data-collection techniques were used including content and discourse analysis of interview data, close examination of Arabic, English and French language sources, both primary and secondary; content analysis of historical records (including newspapers, government documents, IS and other insurgent organisational documents, and autobiographies); and systematic consultation of relevant secondary sources.

Quantitively, the author established the 'Islamic State Ways of Warfare' Database (ISWD) for this book and other projects.[119] Out of its 25 datasets that covers categories of tactics employed by IS in different battlefronts, 8 datasets (see Tables 2.2, 2.3, 2.4, 3.1, 4.1, 4.2, 5.1 and 5.2) were used throughout the following chapters of this book. These datasets are not intended to provide an exhaustive list of categories of tactics employed by IS in specific battles or battlefronts. They rather represent conservative quantitative and qualitative samples of categories of the tactics on which IS relied during combat. They are proxies of the organisation's combat capabilities, tactical innovations and strategic shifts, more than the exact number of times they used these categories of tactics during battles.

Caveats and limitations

A few methodological notes should be mentioned here. First, IS-claimed military operations listed in the book have been cross-checked with media outlets, government sources and/or local sources. Unsurprisingly, there were variations and some of them were adjusted accordingly, mostly to the minimum. Consistent with earlier studies,[120] the majority of IS operations could be confirmed. One of the reasons for the aforenoted variations is that ISI/ISIS/IS definition of an 'armed operation' remains unclear. Some of the operations listed in their military metrics include more than one armed attack. For example, the organisation would list an operation involving an attack by a guerrilla formation (GF), then mortar attacks and then sniping in nearby areas as one single operation, as opposed to three attacks. Second, for an unclear reason(s), the documentation of operations and categories of tactics of IS Provinces in Iraq and Egypt were more consistent than IS Provinces in Syria and Libya. An example would be the comparative figures of 'suicide vehicle-borne improvised explosive devices' (SVBIEDs) in Mosul and Raqqa. IS released over 160 videos or pictures showing SVBIED attacks in Mosul, compared to fewer than 30 for Raqqa. Sinai Province (SP) in Egypt has meticulously reported on its claimed operations as well. Third, the organisation has a consistent tendency to inflate the numbers of its victims while deflating the numbers of certain operations during battles. A clear example of that behaviour was the released figures of the detonated SVBIEDs during the battle of Mosul between 19 October 2016 and 12 July 2017. According to IS' A'maq Agency, the organisation's fighters carried out 482 SVBIED attacks. However, according to one senior officer in the United States Central Command (CENTCOM),[121] IS forces in Mosul detonated over 600 SVBIEDs and large SVBIEDs (SLVBIEDs) between October and December 2016 during the battle of Mosul.[122] This estimate suggests that the overall figure is higher than that claimed by IS. Regardless of the exact figure, IS had the military capacity and the knowhow to detonate at least 482 SVBIEDs in a nine-month and four days series of battles; a capacity that no other ANSA has had up to the date of writing of this book. Qualitatively, the organisation had shown the capacity to innovatively turn an urban terrorism category of tactics into an effective battlefield one.[123]

A final note is due about what this book is *not* about. The book does not aim to provide an exhaustive account of IS battles and battlefronts. It analyses a sample of a high-intensity series of battles, which reflect the level of combat-effectiveness that IS (and potentially other ANSAs) has reached. Nor is it a complete military-historical account of IS battles, although it dedicates several sections to analysing how the organisation developed its military capacities in the four countries. The book does not address the ideological extremism of the organisation, which exceeds the violent extremism and the brutality of al-Qaida and its franchises and is perhaps equivalent to the Saudi *Ikhwan Min Ta'allah* tribal militias of the early twentieth century. The book does not analyse the brutal and regressive behaviour of IS and other well-known features of the organisation, unless they have a significant impact on its military performance. The book is not a historical account of IS and its local governing bodies. It does briefly analyse the socio-political backgrounds in which IS Provinces operate, if relevant to their military build-up processes in the four countries. Finally, the book is neither another critique of counterinsurgency policies nor another counterterrorism manual. It does analyse *some* of the counterinsurgency tactics, it has *several* policy implications and it may forecast how ANSAs – sponsored or unsponsored by states – *will* fight in the future. But the main aim of the book is not to analyse the counterinsurgency policies, but the insurgents' tactics, combat effectiveness and how they fight.

Book structure

The following chapters demonstrate how IS fights and how its earlier expansion and relative endurance deviate from the explanations in the existing literature. They qualitatively test the aforementioned two hypotheses.

Chapter 2 provides a historical overview of the military rise of IS and its predecessors in Iraq. It aims to understand how IS and its predecessors (Islamic State in Iraq and Sham or ISIS; Islamic State in Iraq or ISI; *Hilf al-Mutaybin* or HM/Coalition of the Good; *Majlis Shura al-Mujahidyin* or MSM/Holy Fighters Consultative Council; al-Qaida in Iraq or AQI; and *al-Tawhid wa al-Jihad* in Iraq or TJI/Monotheism and Struggle) were able to gradually develop their combat capacities in Iraq between 2003 and 2017; and as a result, IS was able to occupy Fallujah (January 2014), Mosul (June 2014), Ramadi (May 2015) and other towns. This chapter particularly focuses on analysing how IS in Iraq fought in the three battlefronts of Fallujah, Mosul and Ramadi and how it fights in the aftermath of its territorial defeats.

Chapter 3 aims to understand how IS managed to militarily control and wield influence in approximately 50 per cent of Syria's territories by mid-2015, in the face of many armed actors and their powerful foreign backers. It briefly overviews the development of IS-related and Jihadist military networks between 2003 and 2013. Then it focuses on the fighting capacities and the military build-up after 8 April 2013, when ISIS was declared. The focus subsequently shifts to analysing the military tactics employed to control the whole of Raqqa Governorate. The chapter particularly focuses on analysing how IS in Syria have fought in the battlefronts of Raqqa and how it fights in the aftermath of Raqqa's liberation.

Chapter 4 explains how IS in Libya has managed to build-up its military capacities and how it was able to occupy parts of Derna and all of Sirte. It gives a brief overview of the development of ISI/ISIS/IS-related networks in Derna. Then it focusses on analysing the military tactics employed by IS in the battlefronts of Derna and Sirte.

Chapter 5 focuses on the case of IS in Sinai (Sinai Province or SP) and its military capacities. It overviews the development of the predecessors of IS in Sinai, including *Ansar Bayt al-Maqdis* (ABM or 'Supporters of Jerusalem') and *al-Tawhid wa al-Jihad* in Sinai (TJS or 'Monotheism and Struggle in Sinai'). Despite being ludicrously outnumbered and outgunned, IS in Sinai and its predecessors have survived almost ten years of sustained counterinsurgency campaigns (2011–2020), some of which bordered on 'scorched earth' tactics since September 2013. This chapter analyses the battle of Sheikh Zuweid in July 2015 and develops an explanation of SP's military endurance.

Finally, the book's findings and conclusions are outlined in Chapter 6. These are presented as a framework for understanding the military expansion and endurance of IS, and possibly of other ANSAs. Given the book's engagement with highly relevant case-studies that have been underexplored from this angle, it often goes against the tide of conventional wisdom regarding insurgencies. The book finally aims to contribute to the understanding of insurgencies' combat effectiveness and to offer insights on how ANSAs *may* or *will* fight in the future.

Notes

1. Jayvee (JV) stands for junior varsity (basketball) team. See: David Remnick, 'Going the Distance: On and Off the Road with Barack Obama', *The New Yorker*, 20/1/2014, Accessed on 15/4/2020, at: https://bit.ly/2K6erQv
2. Azmi Bishara, *The State Organisation (Acronym ISIS): A General Framework and a Contribution to Help Understand the Phenomenon*, [in Arabic] vol. 1 (Doha: the Arab Centre for Research and Policy Studies), 2018, p. 155.
3. Tyson was referring to a boxing-bout between legendary boxing champion Floyd Mayweather and mixed martial arts champion Connor McGregor. It

was the first-ever boxing-bout for the latter. He surprisingly endured and fought-back all ten rounds.

4. Jim Scuitto et al. 'ISIS can "muster" between 20,000 and 31,500 fighters, CIA says', *CNN*, 12/9/2014, Accessed on 18/7/2018, at: https://cnn.it/2V4LNWm Other estimates will be provided in the following chapters.

5. For the details of that battle, see the following chapter.

6. Tim Hume, 'Battle for Mosul: How ISIS Is Fighting Back', *CNN*, 25/10/2016, Accessed on 18/7/2018, at: https://cnn.it/34Br2Vv

7. The *Sham* is equivalent to the narrowest possible 'Levant' in historical sense: Syria, Jordan, Israel/Palestine, and Lebanon. In other words, it is equivalent to 'Greater Syria'.

8. See Chapter 3 for a detailed analysis of ISIS/IS' campaigns in Syria.

9. See Chapter 4 for a detailed analysis of IS campaigns in Libya.

10. IS-controlled neighbourhoods in the town-centre of Derna were lost to a coalition of armed nonstate actors led by the *Shura* (Consultative) Council of the Mujahidin and the Revolutionaries of Derna and Its Environs (DMSC for simplification) by July 2015. IS completely withdrew from the suburbs of Derna by April 2016. All of the city of Sirte was liberated from IS forces by December 2016. Cyrenaica Province (*Wilayat Barqa*) lost all of the streets and alleys it controlled in Benghazi to a coalition of militias calling themselves the 'Libyan National Army (LNA)' by February 2017.

11. See Chapter 5 for a detailed analysis of the IS campaigns in Sinai.

12. The book will not be analysing these dimensions, except in terms of their military value for the organisation. Other books and studies have further explored these dimensions. For a comprehensive history of IS and its predecessors and an analysis of its rise as a regional, socio-political phenomenon see: Azmi Bishara, *The State Organisation (Acronym ISIS): A General Framework and a Contribution to Help Understand the Phenomenon*, [in Arabic] vol. 1 (Doha: the Arab Centre for Research and Policy Studies, 2018) and Azmi Bishara (editor), *The State Organisation (Acronym ISIS): Foundation, Discourse and Practice*, vol. 2 [in Arabic] (Doha: Arab Centre for Research and Policy Studies, 2018). On the ultra-extremist, classical Wahhabist ideology of IS see Falih Abduljabbar, *The Caliphate State: Advancing Towards the Past* [in Arabic] (Doha: Arab Centre for Research and Policy Studies, 2017) and Muhammad Abu Rumman and Hasan Abu Haniyah, *The Islamic State: the Sunni Crisis and the Struggle for Global Jihadism* [in Arabic] (Amman: Friedrich-Ebert, 2015); See also, *The ISIS Apocalypse* by William McCants (Picador, 2016) and *ISIS: A History* by Fawaz Gerges (Princeton: Princeton University Press, 2016). On the mass-murders and possible crimes against humanity committed by the organisation see *Inside the Army of Terror*, by Michael Weiss and Hassan Hassan (New York: Regan Arts, 2015); *ISIS: The State of Terror* by J. M. Berger and Jessica Stern (New York: Ecco Press, 2015) and *Black Flags: The Rise of ISIS*, by Joby Warrick (London: Corgi, 2016). On the terrorism tactics employed by the organisation in Europe see *Islamist Terrorism in Europe: A History* by Petter Nesser (Oxford: Oxford University Press, 2018) and Thomas Hegghammer and Petter Nesser, 'Assessing the Islamic State

Commitment to Attacking the West', *Perspectives on Terrorism*, vol. 9, no. 4 (2015). On the sectarian narratives of the organisation see for example Hassan Hassan, 'The Sectarianism of the Islamic State', *Carnegie Papers* (June 2016). On foreign fighters, see Richard Barrett, 'Beyond the Caliphate: Foreign Fighters and Threat of Returnees', *Soufan Center* (October 2007), pp. 14–18. For up-to-date analyses and insights on the organisation, see for example *jihadology.net* and *aymennjawad.org*.

13. Lieutenant-General Terry Wolff (Retd), conversation with author, Prague, 10 February 2017; Lieutenant-General Sean MacFarland, 'Department of Defense Press Briefing by Lieutenant-General Sean MacFarland, Commander, Combined Joint Task Force-Operation Inherent Resolve via Teleconference from Baghdad, Iraq', *Defence.gov*, 10 August 2016, at: https://www.defense.gov/Newsroom/Transcripts/Transcript/Article/911009/department-of-defense-press-briefing-by-lieutenant-general-sean-macfarland-comm/

14. Kathryn M. Cochran and Stephen B. Long, 'Measuring Military Effectiveness: Calculating Casualty Loss-Exchange Ratios for Multilateral Wars, 1816–1990', *International Interactions*, vol. 43, no. 6 (2017), pp. 1024–6.

15. 'The Taliban Are Forced Out of Afghanistan', *BBC History*, 18/7/2018, Accessed on 18/7/2018, at: http://www.bbc.co.uk/history/events/the_taliban_are_forced_out_of_afghanistan

16. 'Baghdad Falls to US Forces', *BBC News*, 9/4/2003, Accessed on 18/7/2018, at: https://bbc.in/34BuDTv

17. From September 2014 (when the US started the attacks before the establishment of the Combined Joint Task Force in October 2014) to July 2017 (when IS lost the City of Mosul).

18. Ben Connable and Martin C. Libicki, *How Insurgencies End* (Arlington: Rand Publications, 2010); Patrick B. Johnston, 'The Geography of Insurgent Organisation and its Consequences for Civil Wars: Evidence from Liberia and Sierra Leone', *Security Studies*, vol. 11 (2008), pp. 107–37; Jason Lyall and Isaiah Wilson, 'Rage against the Machines: Explaining Outcomes in Counterinsurgency Wars', *International Organisation*, vol. 63, no. 1 (2009), pp. 67–106; Andrew Mack, 'Why Big Nations Lose Small Wars: The Politics of Asymmetric Conflict', *World Politics*, vol. 27, no. 2 (1975), pp. 175–200; Sebastian Schutte, 'Geography, Outcome, and Casualties: A Unified Model of Insurgency', *Journal of Conflict Resolution* (March 2014), pp. 1–28; Ivan M. Arreguín-Toft, "How the Weak Win Wars: A Theory of Asymmetric Conflict', *International Security*, vol. 26, no. 1 (2001), pp. 93–128.

19. Arreguín-Toft's study (2001) covers both weaker armed state actors (employing guerrilla and conventional tactics), as well as armed nonstate actors (ANSAs) challenging stronger armed state actors (ASAs).

20. Arreguín-Toft, 'How the Weak Win Wars: A Theory of Asymmetric Conflict', p. 97.

21. Jason Lyall and Isaiah Wilson, 'Rage against the Machines: Explaining Outcomes in Counterinsurgency Wars', *International Organisation*, vol. 63, no. 1 (2009), pp. 67–106.

22. Ibid.
23. The armed conflict is still ongoing in the remaining 16 cases; see: Connable and Martin C. Libicki, *How Insurgencies End*, p. 5.
24. Seth Jones, *Waging Insurgent Warfare* (Oxford: Oxford University Press, 2017), p. 9.
25. Mao Tse-tung, *On Guerrilla Warfare* (Champaign: University of Illinois, [1937] 1961), p. 37.
26. David Petraeus, James F. Amos and John A. Nagl, *The U.S. Army/Marine Corps Counterinsurgency Field Manual* (Chicago: University of Chicago Press, 2007), pp. 79–136.
27. Robert McColl, 'The Insurgent State: Territorial Bases of Revolution', *Annals of the Association of American Geographers* vol. 59, no. 4 (1969), p. 630.
28. Edward G. Lansdale, 'Vietnam: Do We Understand Revolution?', *Foreign Affairs* (October 1964), p. 78. See also: Edward G. Lansdale, *In the Midst of Wars: An American's Mission to Southeast Asia* (New York: Harper & Row, 1972).
29. Bob Dreyfuss, 'How the US War in Afghanistan Fuelled the Taliban Insurgency', *The Nation*, 18/9/2013 Accessed on 18/7/2018, at: https://bit.ly/2yip3sO
30. David Kilcullen, *The Accidental Guerrilla: Fighting Small Wars in the Midst of a Big One* (Oxford: Oxford University Press, 2009).
31. Barbara Walter, 'Why Bad Governance Lead to Repeat Civil Wars', *Journal of Conflict Resolution*, vol. 59, no. 7 (2014), pp. 1242–72.
32. Lawrence Freedman, *Strategy: A History* (Oxford: Oxford University Press, 2013), p. 188.
33. T. E. Lawrence, 'Science of Guerrilla Warfare' (1929) in: Malcolm Brown (ed.), *T. E. Lawrence in War and Peace: An Anthology of the Military Writings of Lawrence of Arabia* (London: Greenhill Books, 2005), p. 284.
34. See: Jones, *Waging Insurgent Warfare*, p. 29; James D. Fearon and David D. Laitin, 'Ethnicity, Insurgency, and Civil War', *American Political Science Review*, vol. 97, no. 1 (2003), pp. 75–6; Doyle Michael W. and Nicholas Sambanis, *Making War and Building Peace* (Princeton: Princeton University Press, 2006), p. 5.
35. See for example: Jones, *Waging Insurgent Warfare*, p. 29. The argument is that al-Qaida failed to initiate an insurgency in Saudi Arabia in 2002. However, there were two low-level insurgencies there between 2003 and 2005 (led by al-Qaida in the Arabian Peninsula – AQAP, with offshoots and ramifications in neighbouring Yemen) and in 2017 and afterwards in the Eastern Province, mainly by Shiite insurgents.
36. Anna Getmansky, 'You Can't Win If You Don't Fight', *Journal of Conflict Resolution*, vol. 7, no. 4 (2012), p. 710. Philip Keefer, 'Insurgency and Credible commitment in Autocracies and Democracies', *The World Bank Economic Review*, vol. 22, no. 1 (2008), pp. 33–61.
37. Michael Engelhardt, 'Democracies, Dictatorships, and Counterinsurgency: Does Regime Type Really Matter?', *Conflict Quarterly*, no. 12

(1992), pp. 52–63; Matthew Wells, 'Casualties, Regime Type and the Outcomes of Wars of Occupation', *Conflict Management and Peace Sciences*, vol. 33, no. 5 (2016), pp. 469–90.

38. This is consistent with other empirical studies comparing different regimes' performance in counterinsurgency. See for example: Engelhardt, 'Democracies'; Alexander Downes, *Targeting Civilians in War* (Ithaca, NY: Cornell University Press, 2008), p. 711; Jason Lyall, "Do Democracies Make Inferior Counterinsurgents? Reassessing Democracy's Impact on War Duration and Outcomes', *International Organization*, vol. 64, no. 1 (2010), pp. 167–92.

39. Downes, *Targeting Civilians in War*, p. 711.

40. Lawrence Freedman, 'Guerrilla Warfare', In: Lawrence Freedman (ed), *Strategy: A History* (Oxford: Oxford University Press, 2013), p. 182.

41. Tse-tung, *On Guerrilla Warfare*, p. 7.

42. Neil Macaulay, 'The Cuban Rebel Army: A Numerical Survey', *The Hispanic American Historical Review*, vol. 58, no. 2 (1978), pp. 284–95; Ernesto Guevara, *Guerrilla Warfare* (North Melbourne: Ocean Press, 1961).

43. Ibid, p. 10.

44. Ibid, p. 21.

45. David Galula, *Counterinsurgency Warfare: Theory and Practice* (Westport, CT: Praeger, 1964), p. 26.

46. McColl, 'The Insurgent State', p. 630.

47. Kenneth E. Boulding, *Conflict and Defense: A General Theory* (New York: Harper, 1962).

48. Sebastian Schutte, 'Geography, Outcome, and Casualties: A Unified Model of Insurgency', *Journal of Conflict Resolution* (March 2014), pp. 1–28.

49. The other three variables are political instability, large population and poverty.

50. Stathis Kalyvas, *The Logic of Violence in Civil War* (New York: Cambridge University Press, 2006).

51. Connable and Libicki, *How Insurgencies End*, pp. 8–9.

52. Great power is determined here based on military strength, economic capability, size of population and territory, resource endowment, and political stability and competence. See: Kenneth N. Waltz, *Theory of International Politics* (New York: McGraw-Hill, 1979), p. 131.

53. Jones, *Waging Insurgent Warfare*, p. 136.

54. See for example: Idean Salehyan, Kristian Skrede Gleditsch, and David E. Cunningham, 'Explaining External Support for Insurgent Groups', *International Organization*, no. 65 (Fall 2011), pp. 709–44.

55. See for example: Steven R. David, 'Why the Third World Matters', *International Security*, vol. 14, no. 1 (Summer 1989), pp. 50–85; Mark P. Lagon, 'The International System and the Reagan Doctrine: Can Realism Explain Aid to 'Freedom Fighters'?' *British Journal of Political Science*, vol. 22, no. 1 (January 1992), pp. 39–70; William Rosenau, 'The Kennedy Administration, US Foreign Internal Security Assistance and the Challenge of "Subterranean War," 1961–63', *Small Wars and Insurgencies*, vol. 14, no. 3 (Autumn 2003), pp. 65–99.

56. See for example: Daniel Byman et al., *Trends in Outside Support for Insurgent Movements* (Santa Monica: RAND Corporation, 2001); Walter Laqueur, *Guerrilla Warfare: A Historical and Critical Study* (New Brunswick: Transaction, 1998).

57. Rob Kevlihan, *Aid, Insurgencies and Conflict Transformation: When Greed is Good* (London: Routledge, 2012), pp. 10–17. Tom Keating, 'The Importance of Financing in Enabling and Sustaining the Conflict in Syria (and Beyond)', *Perspectives on Terrorism*, vol. 8, no. 4 (2014), Accessed on 18/7/2018, at: https://bit.ly/2xkON85

58. See for example: Anthony Cordesman, *Arab-Israeli Military Forces in an Era of Asymmetric Warfare* (London: Praeger, 2006), pp. 254–61.

59. Freedman, *Strategy: A History*, p. 189.

60. Ibid, p. 189.

61. Stathis N. Kalyvas and Laia Balcells, 'International System and Technologies of Rebellion: How the End of the Cold War Shaped Internal Conflict', *American Political Science Review*, vol. 104, no. 3 (August 2010), p. 427.

62. The Arabic acronym for the 'Islamic State in Iraq and Sham' (ISIS). It is usually used as a derogative term to describe IS and its predecessors or violent extremist ideas, practices, organisations or individuals.

63. 'Partners Archive', *The Global Coalition against Daesh*, Accessed on 19/7/2018, at: https://bit.ly/34CjRME

64. Lyall and Wilson, 'Rage against the Machines', pp. 67–106.

65. Hans Seig, 'How the Transformation of Military Power Leads to Increasing Asymmetries in Warfare?', *Armed Forces and Society*, vol. 40, no. 2 (2014), pp. 332–56; Seth Jones and Patrick Johnston, 'The Future of Insurgency', *Studies in Conflict and Terrorism*, vol. 36, no. 1 (2013), pp. 1–25; David Kilcullen, *Out of the Mountains: The Coming Age of the Urban Guerrilla* (New York: Oxford University Press, 2013).

66. Engelhardt, 'Democracies', p. 52; on the effectiveness of terrorism see the seminal work of Bruce Hoffman, *Inside Terrorism* (New York: Columbia University Press, 2006).

67. Alberto Bayo, *150 Questions for a Guerrilla* (Denver: Cypress, 1963), p. 12.

68. Engelhardt, 'Democracies', p. 52.

69. Jones, *Waging Insurgent Warfare*, p. 59.

70. See for example: Bruce Hoffman, *Inside Terrorism*, revised ed. (New York: Columbia University Press, 2006), pp. 131–71; Robert A. Pape, *Dying to Win: The Strategic Logic of Suicide Terrorism* (New York: Random House, 2005); Robert A. Pape and James K. Feldman, *Cutting the Fuse: The Explosion of Global Suicide Terrorism and How to Stop It* (Chicago: University of Chicago Press, 2010).

71. Jones, *Waging Insurgent Warfare*, p. 74.

72. Especially against the Japanese forces in the Battle of Tai'erzhuang between March and April 1938.

73. 'Afghanistan: US and Taliban sign historic peace deal hailed as "momentous day"', *Sky News*, 29/02/2020, at: https://news.sky.com/story/

us-and-taliban-sign-historic-peace-agreement-hailed-as-momentous-day-11946209. For a detailed analysis of the process, see Omar Ashour (ed.), *Bullets to Ballots: Collective De-Radicalisation of Armed Movements* (Edinburgh: Edinburgh University Press, forthcoming 2021); see also: Jonathan Powell, *Talking to Terrorists* (London: Vintage, 2015).

74. Arreguín-Toft, *How the Weak*, p. 100; p. 122.

75. Ibid, p. 111.

76. Engelhardt, 'Democracies', p. 52.

77. See for example: Arreguín-Toft, *How the Weak*, pp. 34–43. For a critique see: Jones, *Waging Insurgent Warfare*, pp. 52–4.

78. Jones, *Waging Insurgent Warfare*, p. 170.

79. See for example: Francisco Gutiérrez Sanín and Elisabeth Jean Wood, 'Ideology in Civil Wars', *Journal of Peace Research*, vol. 51, no. 2 (2014), pp. 213–26.

80. Omar Ashour, 'Ballots to Bullets: Transformations from Armed to Unarmed Activism', *Strategic Papers*, Arab Centre for Research and Policy Studies (2019), Accessed on 15/4/2020, at: https://bit.ly/2wKGNNk

81. Tse-tung, *Guerrilla Warfare*, p. 18.

82. Omar Ashour, *The De-Radicalization of Jihadists: Transforming Armed Islamists Movements* (London: Routledge, 2009), p. 138.

83. See for example: Paul Staniland, 'Organizing Insurgency: Networks, Resources, and Rebellion in South Asia', *International Security*, vol. 37, no. 1 (2012); Douglas Pike, *Viet Cong: The Organization and Technique of the National Liberation Front of South Vietnam* (Boston: MIT Press, 1966).

84. Kenneth Pollack, *Armies of Sands* (Oxford: Oxford University Press, 2019), p. 996.

85. For levels of support see for example The Arab Opinion Index survey on ISIL, in: "The 2017–2018 Arab Opinion Index: Main Results in Brief," *Arab Centre for Research and Policy Studies*, at: https://bit.ly/38xi9OF, pp. 34–36.

86. Ibid, pp. 1009–21.

87. Although, if its foes are *weak* by Western military standards (and this is a fair conclusion in the author's opinion), IS was certainly *weaker* as explained in the following chapters.

88. See for example the observations of Abdul Haqq Layada, the first *emir* of the GIA, on the zeal and other features of the GIA's company-sized 'Green Battalion' (also known as 'the Death Battalion') operating in Tala'cha, Chrea Mountains in Blida Province (in Abdul Haqq Layada, 'Layada to *al-Hayat*: I Founded the Armed Group to Defend the Algerian People and Not to Kill them' [in Arabic], Interview by Camille al-Tawil, *al-Hayat*, 6 June 2007, p. 6); see also Umar Shaikhy's observations in Umar Shaikhy, 'The *Emir* of al-Akhdariya Uncovers the Secrets of the Mountain' [in Arabic], Interview by Camille al-Tawil, *al-Hayat*, 7 June 2007, p. 10.

89. See for example the observations of Fathy Bin Suleiman (Abdul Rahman al-Hattab), one of the military commanders and consultative council members of the Libyan Islamic Fighting Group (an Afghanistan war veteran who fought in Algeria among other experienced foreign fighters) on the combat capacities, manpower and armament of the GIA in Camille al-Tawil, *al-Qaida and Its Sisters* [in Arabic] (Beirut: Dar al-Saqi, 2007), pp. 127–9; Noman Benotman, Interview by author, London, 11 June 2009. I should note that the cases of the Afghan-trained (and the Hizbullah-trained), non-Algerian fighters in the Algerian insurgency and their impact on the GIA's combat performance are complicated. Due to space limitations and thematic considerations, I will not be able to fully examine the issue in this book.

90. On empirical observations and accounts relevant to these variables, see for example: Abu Mus'ab al-Suri, *The Summary of My Testimony on the Jihad in Algeria* [in Arabic] (Afghanistan: [No Publisher], [No Date]); Habib Souaïdia, *La Sale Guerre* (Paris, éditions La Découverte, 2001); Camille al-Tawil, *al-Qaida and Its Sisters* [in Arabic] (Beirut: Dar al-Saqi, 2007).

91. For a background see for example: Omar Ashour, *The De-Radicalization of Jihadists: Transforming Armed Islamists Movements* (London: Routledge, 2009), pp. 56–62, 110–20, 127–34.

92. See for example the other seminal work of Kenneth Pollack: Kenneth Pollack, *Arabs at War: Military Effectiveness, 1948–1991* (Lincoln: University of Nebraska Press, 2004), pp. 100–58.

93. Arreguín-Toft, *How the Weak*, p. 103.

94. Edward Luttwak, 'The Operational-Level of War', *International Security*, vol. 5, no. 3 (Winter 1980–81), pp. 61–79.

95. Omar Ashour (ed.), *Punching Above Weights: Combat Effectiveness of Armed Non-State Actors* (Doha: Arab Centre for Research and Policy Studies) (forthcoming, 2021). See also: the proceedings of the annual conference of the Strategic Studies Unit entitled 'Militias and Armies: Developments of Combat Capacities of Armed Non-State and State Actors', Arab Centre for Research and Policy Studies (ACRPS), Doha, 24/2/2020, at: https://bit.ly/2K3hYzc

96. For alternative definitions and related explanations, see for example: Allan Millett et al. 'The Effectiveness of Military Organizations', *International Security*, vol. 11, no. 1 (1986): pp. 37–71; Stephen Biddle, *Military Power: Explaining Victory and Defeat in Modern Battle* (Princeton: Princeton University Press, 2006); Risa Brooks, 'Making Military Might: Why Do States Fail and Succeed?', *International Security*, vol. 28, no. 2 (2003): pp. 149–91.

97. See: Omar Ashour, 'Post-Jihadism and Ideological De-Radicalization', in: Zaheer Kazimi and Jeevan Doel (ed), *Contextualizing Jihadi Ideologies* (New York: Columbia University Press, 2012).

98. See for example: Madawi al-Rasheed, *A History of Saudi Arabia* (Cambridge: Cambridge University Press, 2010), pp. 57–66.

99. Ibid, pp. 46–7.

100. For example, the Algerian Armed Islamic Group (GIA) declared the rest of the Algerian population apostates in 1997. See: Ashour, *The De-Radicalization of Jihadists*, p. 59.

101. See for example: Mona El-Tahawy, 'Lives Torn Apart', *The Guardian*, 20/10/1999 on the case of Nasr Abu Zeid who was declared an 'apostate' by al-Azhar, the official religious establishment in Egypt.

102. The case of the post-1995 GIA is a prime example. See: Ashour, *The De-Radicalization of Jihadists*, pp. 57–9. It should be noted that the two terms 'Takfirist' and 'Wahhabist' have negative connotations in the Arabic language. Both are widely used in propaganda to smear political, sectarian and religious opponents. Needles to mention, this is not the intention here and hence the definitions are provided to clarify the meanings in this book.

103. Al-Rasheed, *A History of Saudi Arabia*, p. 57; Bader Al-Ibrahim, 'ISIS, Wahhabism and Takfir', *Contemporary Arab Affairs*, vol. 8, no. 3 (2015), pp. 408–15.

104. That said, there is still an ongoing debate on how to categorise IS and its predecessors, mainly either under a Salafi-Jihadist or a Violent Takfirist category.

105. Muayyed Bajis, 'IS Executes Sharia Judge for Extreme Takfirism', [in Arabic] *Arabi21*, 8/8/2015, Accessed on 15/4/2020, at: https://bit.ly/2z3d0jI

106. For a sample of these studies, see footnote no. 12, this chapter.

107. This is not a comprehensive analysis of the individual members of IS. This is a brief typology of fighters affiliated with the organisation, whose contributions matter on the battlefield.

108. A.W. Former FSA Commander in Raqqa, Interview by author, Istanbul, 9 October 2016.

109. Such as limited manpower, firepower, financial ability, capacity to control or deny territory and other limitations common among armed non-state actors.

110. For a region-wide analysis see: Jean-Pierre Filiu, *From Deep State to Islamic State: The Arab Counter-Revolution and Its Jihadist Legacy* (New York: Oxford University Press, 2005).

111. See for example: Ashour, 'Post-Jihadism and Ideological De-Radicalization'.

112. See for example: Mohammed Hafez, *Why Muslims Rebel? Repression and Resistance in the Islamic World* (London: Lynne Rienner Publishers, 2003).

113. A governorate (*muhafaza*) is an administrative division in Iraq, Syria and Egypt; a district is one possible translation for the term *sha'biyya* used in Libya under Gaddafi.

114. As opposed to Sirte, Sheikh Zuweid had low-to-no strategic value for IS. However, the level of investment in the attack to occupy Sheikh Zuweid in July 2015 was high. Hence, the attack reflected the tactical and operational capabilities of Sinai Province.

115. Ryan Dillon and Eric J. Pahon, 'Department of Defense Press Briefing by Colonel Dillon via Teleconference from Baghdad, Iraq', *US Department of Defense*, 7/10/2017.
116. For more details, see the following four chapters.
117. The author is grateful for Aaron Zelin and the *Jihadology* website for the help in accessing many of ISI/ISIS/IS publications, audio- and video-releases.
118. As aforenoted, IS definition of an 'armed operation' is unclear. Some of the operations listed in their databases include more than one attack.
119. The author is grateful to his research team at the Arab Centre's Strategic Studies Unit (SSU) and the Doha Institute's Critical Security Studies Graduate Programme (MCSS). The database and many of the data-collection activities (including interviews) were made possible by the support of the Doha Institute for Graduate Studies Major Research Fund grant (MRF 03-01). The findings achieved herein are solely the responsibility of the author. The author is also grateful to the Arab Centre for Research and Policy Studies for the support throughout this research project.
120. Daniel Milton, Bryan Price and Muhammad al-'Ubaydi, 'The Islamic State in Iraq and the Levant: More than Just a June Surprise', *CTC Sentinel*, vol. 7, issue 6 (June 2014), p. 2. Their sample included 345 claimed attacks and was geographically limited to Baghdad. 57 per cent of the attacks (198 claimed attacks) were confirmed by local or international media. However, 85 per cent of the unreported attacks in the media yielded less than three casualties. Hence, they were of limited impact to attract the media attention, but they were still documented and reported by ISI.
121. The United States Central Command is one of the eleven unified combatant commands of the US Department of Defense (including the six regional combatant commands). It was established in 1983 and its 'Area of Responsibility' includes the Middle East and Central Asia, most notably Iraq and Afghanistan. As of 2014, CENTCOM forces are deployed in Iraq and Syria as part of Operation Inherent Resolve.
122. Mark Perry, 'How Iraq's Army Could Beat ISIS in Mosul, But Lose the Country', *Politico*, 15/12/2016, Accessed on 15/4/2020, at: https://politi.co/38r6v65
123. For more details, see the following two chapters.

2

Implodes but Expands: How the 'Islamic State' Fights in Iraq

In the break time, bury the TNT and bring the dollar.

A common phrase within IS ranks

It is extraordinary. They were outgunned, outnumbered [during] the whole thing from start to finish . . ., yet they still got all this done.

Colonel Pat Work, Commander of the 2nd Brigade
Combat Team during the battle for liberating
Mosul from ISIS[1]

ISIS [weapon-manufacturing] is unlike anything we have ever seen from a non-state force.

Solomon H. Black, Weapons Analysis Specialist
in the US State Department[2]

A cradle, but not yet a grave

This chapter overviews the military development of IS and its predecessors in Iraq. It aims to understand how IS and its six predecessors (ISIS, ISI, HM, MSM, AQI and TJI)[3] were able to gradually develop their combat capacities in that country from 2003 onwards. As a result, IS was able to occupy Fallujah in January 2014, Mosul in June 2014, Ramadi in May 2015 and other Iraqi towns. The chapter focuses on analysing the three battlefronts of Fallujah, Mosul and Ramadi, as a sample reflecting how ISIS/IS have fought in Iraq. The chapter also analyses how IS fought after the successful liberation efforts by the Iraqi army and its allies, up to March 2020. The chapter is partly based on interviews and documents produced by the Iraqi government and documents of ISIS/IS in Iraq some of which were confiscated and then released by the US government, and other open-source material.

The chapter is composed of seven sections. The following section gives an overview of the military build-up of IS and its predecessors in Iraq from October 2002 – the date when the founding leader of its predecessor organisations, Abu Mus'ab al-Zarqawi,[4] arrived in the northeast of the country – to June 2014, the date of IS establishment and its self declaration as a 'caliphate'.[5] The third, fourth and fifth sections outline the details of the three abovementioned battlefronts within specific timeframes.[6] The sixth section of the chapter analyses how IS fights in Iraq, using data and observations from the three battlefronts and elsewhere in Iraq. Finally, the concluding section focuses on the future of IS insurgency in Iraq, after losing territory and shifting back to guerrilla and terrorism strategies and tactics.

Striking from the ashes: an overview of IS military build-up in Iraq

By June 2014, IS had already captured Fallujah (Anbar Governorate), parts of Ramadi (Anbar Governorate), Mosul (Nineveh Governorate), Tikrit (Salahuddin Governorate) and several smaller towns and villages in Iraq. This was a direct result of a significant military build-up, tactical innovations in urban terrorism and shifts between conventional and guerrilla warfare. By 2016, IS had enough resources to carry out a staggering 1,112 suicide bombings; 842 of which were SVBIEDs.[7] The Nineveh and Anbar 'Provinces' of IS dominated the metrics, with 311 and 193 suicide operations respectively. That capacity is non-existent under any other armed non-state actor in the modern world. The predecessors of IS/ISIS/ISI also lacked such a capacity. Al-Zarqawi's words in 2005 highlight the gap between the resources of *al-Tawhid wa'l-Jihad* in Iraq (TJI) and those of IS:

> there can be no comparison between our capabilities and the enemy's resources. Hundreds of our brothers are fighting hundreds of thousands of the enemy. As to military equipment . . . the brothers possess their firearms, some mortars and some rocket-propelled grenades against tanks, armoured vehicles and aircraft.[8]

Al-Zarqawi was also complaining about the lack of supportive geography and 'strategic depth'[9] in Iraq. 'The land of jihad in Iraq is different from Afghanistan and Chechnya . . . it is [flat] like a hand.

There are no mountains, valleys or jungles', he said in an interview.[10] Hence, IS/ISIS/ISI's predecessors suffered from structural, strategic, operational and tactical weaknesses including the lack of supportive geography, lack of popular support, limited capacities and training in IED-manufacturing and mediocre skills in close-quarter combat, small-unit firefighting and marksmanship. The organisation also suffered from limited knowledge of heavy-weapon handling, especially armoured and artillery, combined arms tactics, a complete lack of industrial scale VBIED-manufacturing and unmanned combat aerial vehicles (UCAVs or combat drones). These knowledge gaps have been steadily filled since the creation of TJI in the fall of 2003. Between 2003 and 2006, the most common tactic used against the American troops by two of IS predecessors, TJI and AQI, was static, covert roadside IEDs. Although Zarqawi clearly stated that these types of IEDs (and mortars) will never 'win the war in Iraq',[11] their impact was still significant. Static, roadside IEDs accounted for 45 per cent of all US fatalities from enemy action between March 2003 and January 2007.[12] By then, the tactical capacities of the organisation had been enhanced significantly.

According to the IS *al-Naba'* newsletter, both AQI and its successor ISI, had developed squad-sized units (called *mafariz*)[13] specialised in sniping, small-unit engagements at short-range and other aspects of guerrilla and terrorism warfare by January 2007.[14] But by December 2007, mainly due to the American 'Surge';[15] the Awakening Councils;[16] infighting between insurgents; and the general public's transsectarian backlash against ISI; the organisation's military capacities had almost been destroyed. 'We now have no place where we can stand for even a quarter of an hour', said the former emir of ISI, Hamed al-Zawi (Abu Omar al-Baghdadi),[17] in early 2008.[18] As a result, Abu Hamza al-Muhajir,[19] the so-called 'defence minister' of ISI, decided to shift strategies and ways of warfare from urban guerrilla-intensive to terrorism-intensive. The new strategy aimed at building up ISI's resources for a sustained urban terrorism campaign. In March 2008, al-Muhajir dissolved all non-IED tactical units, including sniping units and guerrilla formations. He focused all ISI resources on building IED-tactical units.[20] These included 'IED-manufacturing' units, 'detonator-manufacturing' units, 'planting, booby-trapping and camouflaging' units and 'detonation-units'.[21] By the end of 2010, ISI had extensively utilised five types of IEDs: remote-controlled IEDs (RCIEDs), booby-trapped IEDs (BTIEDs) known within ISI as 'traps of fools',[22] remote-controlled booby-trapped IEDs (CBTIEDs), static-covert IEDs/mines (planted on roadsides and open fields) and VBIEDs. However, IED-intensive tactics were more lethal in killing and terrorising civilians as opposed to effectively fighting the regular forces (as Zarqawi concluded earlier in a 2006 interview with AQI's media office). Between 2003 and 2010, suicide bombings using IEDs have killed twelve thousand civilians and two hundred soldiers in the American-led coalition. The overwhelming majority of the fallen soldiers (175) were Americans.[23]

If ISI was able to innovate, manufacture and weaponise IEDs after the Surge during its weakest phase (2008–10), it did much more than urban terrorism and killing civilians after 2010. Between November 2011 and April 2014, the organisation and its allies captured weapons from, and looted depots of, both Iraqi and Syrian state and nonstate armed actors. ISI/ISIS/IS also started designing their own munitions and manufactured some of them during this period and afterwards. The annual military metrics reports during the Islamic Lunar (*Hijri*) years of 1433 (November 2011 to November 2012) and 1434 (November 2012 to November 2013) reflected the new capacities.[24] This is in addition to the ISIS monthly military-metrics reports published by the organisation's older version of *al-Naba'* newsletter (issues 50, 51, 52 and 53) between November 2013 and April 2014.[25] During this period – just before the establishment of IS in June 2014 – ISI/ISIS allegedly executed 16,603 operations in northern and central Iraq, with a monthly average of 553 armed operations.[26] The largest percentage of these operations (41 per cent) were claimed in the Nineveh governorate, whose provincial capital is Mosul. The rest were claimed in the Salahuddin governorate whose provincial capital is Tikrit (21 per cent), the Anbar governorate whose provincial capital is Ramadi (19 per cent), Diyala whose provincial capital is Ba'qubah (13 per cent) and Baghdad (6 per cent).[27] Those attacks involved the extensive use of IEDs, VBIEDs/MBIEDs,[28] SVBIEDs/SMBIEDs,[29] booby-traps, suicide vests (SVs), guerrilla formations (GFs), suicidal guerrilla formations (SGFs or *inghimassiyun*), light and medium artillery, guided and unguided rockets and sniping.

At this phase of development, ISI/ISIS were still behind the more developed phases of IED mass-production schemes and advanced weapon-manufacturing techniques. This was developed from late 2014 up to late 2017, when several weapon components were standardised such as locally manufactured injection-moulded munition fuses, shoulder-fired rockets, mortar ammunition, modular bomb parts and plastic-bodied landmines that underwent generations of upgrades. These and other weapons and ammunitions were produced in industrial quantities,[30] before such capacities were degraded or majorly destroyed by the American-led Combined Joint Task Force (CJTF) during Operation Inherent Resolve (OIR).[31]

During the 2014–17 phase, IS was gradually building a system of steady procurement, armaments production that combined research and development, mass-production and systematic distribution of products and knowhows to amplify the organisation's endurance and power, as well as to enhance the combat skills of its members.[32] By early 2015, IS possessed an arsenal that could be utilised in conventional, guerrilla and terrorism operations. This included armoured warfare pieces such as the American-made M1 Abrams tanks, the Soviet-made T-72, T-62 and T-55 and BM1 and BM2 Soviet-made infantry fighting vehicles, among others.[33] Artillery pieces were also a part of IS' conventional military build-up during that phase. They included pieces as heavy as the American-made 155mm towed M198 howitzers[34] and the Chinese-made 133mm towed Type 58 gun.[35] In air defence, IS had an out-of-date conventional arsenal that included old ZU-23 anti-aircraft autocannons. More dangerous in the air defence arsenal were an unknown number of FIM-92 Stingers,[36] SA-16[37] and SA-18[38] guided surface-to-air missiles (man-portable air-defence systems or MANPADS). The American-made Stingers were looted from the Iraqi army depots in Mosul. Some of the Soviet-made SA-16 and SA-18 were taken from the Syrian army depots in Tabqa military airbase in Raqqa Governorate.[39] Also, IS had an unconfirmed number of Soviet-made SA-7 guided surface-to-air missiles.[40]

In addition to conventional capacities, by late 2014 IS provinces in Iraq had one of the most complex guerrilla arsenals of any armed nonstate actor in the region and the world. As a reflection of their guerrilla capacities, Conflict Armament Research (CAR) has documented 1,270 weapons and 29,168 units of ammunition that belonged to IS in Iraq alone between July 2014 and November 2017.[41] This sample featured 31 types of weapons, all of which are commonly used in guerrilla tactics with the exception of (a less common) 120mm heavy mortars.[42]

When it comes to terrorism capacities, IS had a decade of experience in IED-manufacturing and utilisation, as explained earlier.

The organisation inherited ISI's experiences in small-scale production of suicide belts and vests for individuals, and of fitting passenger cars and motorcycles with explosives to produce S/VBIEDs and S/MBIEDS in small workshops. It built on these experiences in terrorism to establish industrial-scale IED production facilities, with the capacity to rig large trucks and captured armoured military vehicles with explosives. The end-products were various types of SVBIEDs and SLVBIEDs.

By early 2015, IS had completed its military build-up. It had transformed from an almost-destroyed group relying overwhelmingly on IED-intensive terrorism and secret cells in 2010, to a transnational organisation with significant conventional, guerrilla and terrorism capacities only four years later. To adapt to a changing environment and to shifting ways of warfare, the military structures of ISI/ISIS/IS kept on rapidly mutating in terms of sizes (ranging from squad-sized *mafariz* to regiment-sized *juyush* or 'armies'),[43] tactical and operational command-and-control/orders of battles, the reduction or expansion of specialisations and missions and even the titles of the military structures and the military-strategic command of the organisation (such as the 'war ministry' in ISI, and the 'military council', the 'supreme war department', and the 'general military office' in ISIS/IS). Four features of this changing organisational structure were both constant, and combat impactful. The first is that regardless of the sizes and the titles, there was a commitment to, and an emphasis on, specialisations and missions. ISI had squad-sized sniping and intelligence *mafariz*. Those became company-sized and battalion-sized formations under IS (*Saraya al-Qans* and *al-Maktab al-'Amni* respectively). The same changes applied to specialised missions. *Mafariz al-Isnad* (support detachments) transformed from squad-sized to battalion-sized units called *al-Quwa al-Markaziya* (central force, colloquially known as *al-Markaziyat* or 'the centrals' in Syria). These units had similar tasks to quick reaction forces (QRFs) and special forces in conventional state armies.

Second, given the constant manpower crisis of IS and its predecessors, the tactical and operational hierarchies were almost always based on need and *'urf* (tradition) as opposed to internal regulations or standard operating procedures (SOPs). For example, a *wali* (governor) of an IS 'Province' would fight as a soldier under the military emir (commander) of the *qati'* (the military sector)[44] during some of the *iqtihamat* (incursions) operations. This is despite that the *wali* of a province is supposed to be hierarchically above the *emir* of a sector. Under IS, the common '*urf* was that if an IS 'Province' is a battlefront or bordering a battlefront, the tactical and operational-level command-and-control is delegated from the military council to the local sectors' commanders of that 'Province'. The hierarchy would be overturned. These tactical and operational-level commanders would be prioritised in terms of rapid deployment and support requests.

Third, IS constantly diversified its combat units by selecting fighters from different geographical sectors for a battle. In the case of annihilation on the battlefield, the losses would be shared among several sectors, instead of being concentrated in one sector. This was done to reduce public backlashes and potential local rebellions against IS governance structures, as well as to address the manpower crisis of the organisation. This had the (likely) unintended ramification of diversifying the experiences and the combat-cultures (or combat-multiculturalism) of the fighters involved in the same operation(s).

A fourth constant feature is about the degree of self-reliance of ISI/ ISIS/IS combat units regardless of their sizes, titles and places in the rapidly changing organisational structures. The degree of self-reliance at the *mafariz*-level was usually high. It sometimes bordered on self-sufficiency. IS fighters had a common saying to reflect that: 'in the break-time, bury the TNT and bring the dollar'.[45] This means in the times of no combat, the fighters should bury or store their arsenals and start fundraising to self-finance operations and arms-purchases to be ready when it is combat-time. This self-reliance feature hails from the days of the TJI, when the predecessor of IS were small terrorist cells with limited resources.

Two final observations on ISI/ISIS should be highlighted here regarding their military build-up, as they also highlight organisational adaptiveness and resilience while building capabilities for committing mass-violence in Iraq. First, ISI/ISIS/IS military build-up had happened despite the sustained decapitation of its 'war ministers' and heads of ISI/ISIS/IS military councils as well as middle-ranking commanders. In May 2010, 80 per cent of ISI's military commanders (thirty-four out of forty-two senior commanders) had been either killed or captured according to General Ray Ordeino, the Head of the Joint Forces Command in Iraq.[46] On a rank-and-file level, Leon Panetta, the US secretary of defence back then, told the American Senate in June 2010 that only one thousand ISI fighters remained active.[47] Overall, the longest serving military commander of ISI/ISIS/IS was Abu

Hamza al-Muhajir. His 'war ministerial' tenure lasted for about three and a half years (October 2006 to April 2010). Nu'man al-Zaidi (Abu Suleiman Nasser al-Din), who replaced him as 'war minister', was killed after only 13 months (in May 2011). Samir al-Khlifawi (Haji Bakr), who replaced al-Zaidi as the military commander and who is credited with further institutionalising, professionalising and enhancing specialisations within the ISI military council, was killed after only two years and eight months (in January 2014). He was succeeded by Adnan al-Bilawi (Abu Abdul Rahman al-Bilawi). The latter was the mastermind of the Mosul takeover in June 2014. He blew himself up to avoid being captured by the Iraqi Federal Police, just six days before ISIS completed its takeover of Mosul (10 June 2014). His tenure lasted for six months. Al-Bilawi was succeed by Adnan al-Suwaydawi (Abu Muhannad al-Suwaydawi or Haji Dawud) who was killed five months later (in November 2014). His successor was Fadel al-Hiyali (Abu Muslim al-Turkmani or Haji Mutaz), who was also killed ten months later in August 2015. Afterwards, the position of the head of the military council[48] was occupied by Iyad al-'Ubaydi (Abu Salih Haifa), who was killed in the final stages of the battle of (West) Mosul.[49] Overall, three heads of the military council were killed by American airstrikes. Four were killed by Iraqi and Syrian ground forces, both state and nonstate ones.

Briefly, Table 2.1 (overleaf) illustrates the sustained decapitation of IS military commanders. It reflects the ability of the organisation to continue growing even when its top military commanders were neutralised.

Second, ISI had a counterplan to compensate for the loss of leadership. This counterplan was a coordinated one-year campaign that started in summer 2012 with the objective of freeing ISI's leading cadres. 'Breaking the Walls', as the campaign was codenamed, was critical for the organisation's military build-up. The campaign resulted in freeing well-experienced senior commanders who led ISI/ISIS/IS' operational objectives, including occupying and governing territories. Overall, 'Breaking the Walls' featured over 900 VBIED-led or VBIED-intensive attacks, over 100 SVBIED-led or SVBIED-intensive attacks and eight attempted or successful prison-breaks across Iraq between July 2012 and July 2013.[50] In total, ISI/ISIS was able to free 824 imprisoned cadres by sheer force; a first even compared to other spectacular prison-breaks such as the Sarposa and the Ghazni ones by the Taliban in 2011 and 2015 respectively, the HM Prison Maze escape by the Provisional Irish Republican

Table 2.1 ISI/ISIS/IS Heads of the Military Council

Commander	Start Date	End Date	Leadership Period	Termination Reason
Abu Hamza Al-Muhajir	October 2006	April 2010	42 Months (3.5 Years)	Killed in action (KIA) (by American–Iraqi Joint Forces)
Abu Suleiman Nasser al-Din	April 2010	May 2011	13 Months	KIA (by Iraqi Forces)
Haji Bakr	May 2011	January 2014	30 Months (2.75 Years)	KIA (by Syrian Opposition Fighters)
Abu Abdul Rahman al-Bilawi	January 2014	June 2014	6 Months	KIA (by Iraqi Forces)
Abu Muhannad al-Suwaydawi	June 2014	November 2014	5 Months	KIA (by a US Airstrike)
Abu Muslim Al-Turkmani	November 2014	August 2015	10 Months	KIA (by a US Airstrike)
Abu Salih Haifa	August 2015	July 2017	23 Months (1.9 Years)	KIA (by Iraqi Forces)

Army in 1983 and the Punta Carretas escape by the Uruguayan Tupamaros in 1971.[51] According to Iraqi security officials, the 842 cadres included 276 of the 'most dangerous' mid-ranking and senior commanders.[52] The campaign was a strong indicator of ISI/ISIS' successful military build-up and capacity to rerise and strike back after a crushing defeat in 2010.

The battlefront of Fallujah: January 2014–June 2016

Fallujah was the first town to fall into ISIS' territories after its resurgence. The final battle to control the town raged between 30 December 2013 and 3 January 2014. On that date, ISIS declared

Fallujah as part of its territories during Friday prayers. However, ISIS only took control of the majority of neighbourhoods by the end of March 2014.[53] Also, ISIS did not fight alone, like the case in some of the other Iraqi cities and towns. In Fallujah specifically, six other armed organisations were in tactical alliance with it: the Islamic Army, the Mujahidin Army, the Twentieth Revolution Brigades, the Army of the Men of the Naqshbandi Order (or the Naqshbandi Army), the Fallujah Military Council and [Abdullah] al-Janabi Group. Between 30 December and 3 January, at least 18 armed formations belonging to the aforementioned organisations were operating in Fallujah and its suburbs against Iraqi regular forces. About one third of these formations belonged to ISIS.[54] In terms of manpower, ISIS formations ranged between 200 and 300 fighters, a small fraction of the manpower of the Iraqi regular forces in and around the town. ISIS alone, without its tactical allies, was outnumbered in Fallujah by an estimated 50-to-1.[55]

By the end of January 2014, ISIS had grown to become the most dominant armed organisation in Fallujah, with hundreds of new recruits and defectors from other organisations. However, ISIS did not fully control the town despite declaring that on 3 January 2014 ISIS largest military parade of about 100 armed vehicles southwest of Fallujah only occurred on 31 March 2014.[56] It included captured Iraqi Army Hummers and an M113 tracked armoured personnel carrier. It was an indication of confidence, control and a demise of anti-ISIS local resistance.[57]

On the offensive: expansion

ISIS' attainment of victory in Fallujah developed via a three-phase, operational-level *modus operandi*: softening and creeping (SC), coalition-building (C) and liquidating-consolidating (LC). Given that this *modus operandi* will be repeated elsewhere, I shall refer to it as SCCLC for simplification (pronounced as 'skulls'). In part or in whole, SCCLC was executed in Iraqi, Syrian, Libyan, Egyptian towns and cities as well as elsewhere, with mixed outcomes.[58] In Fallujah, ISIS wore down the Iraqi military and security forces by creeping back into its former stronghold.[59] It raised the number of operations in Fallujah from an average of one-armed operation every two days in the first quarter of 2013, to double that (one operation every day) in the last quarter of the same year. However, this phase involved more than just quantitative intensity. Qualitatively, ISIS used surgical strikes to liquidate selected individuals. It shunned mass-murder terror tactics like those used elsewhere in Iraq and the world during that phase to avoid local backlashes and infighting with other armed groups. The surgical tactics included close-quarter assassinations, under-vehicle IED bombings and various punishment methods, such as using IEDs and VBIEDs to blow up the houses of selected anti-ISIS individuals including local tribal leaders and police commanders.[60] ISIS also selected specific businessmen to extort via terrorising them as well as assassinating anti-ISIS Sunni clerics. Around Fallujah, 'qualitative' operations included the killing of army commanders and gradually disrupting, harassing and/or cutting off the Iraqi army's supply lines. Assassination victims included General Mohammed Ahmed al-Kurawi, the Commander of the Seventh Division in the Iraqi Army. He was involved in an earlier bloody crackdown on a sit-in protest in Hawijah in April 2013.[61] General al-Kurawi was operating in the Anbar Governorate when he was killed by suicide-bombers wearing SVs.[62] ISIS units also managed to isolate several neighbourhoods in Fallujah, Ramadi and other parts of the Anbar Governorate by blowing up bridges and mining roads. Overall, 85 bridges were destroyed by ISIS/IS in the Anbar Governorate with IEDs and VBIEDs between early 2014 and mid-2015.[63]

A second operational-level phase – coalition-building – has been taking place since mid-2013, at the latest. ISIS aimed to neutralise and reverse the *Sahawat*'s (Awakening Councils) impact. Rather than engaging in hostilities, ISIS selectively formed a series of tactical and operational alliances and pursued pragmatic coalition-building with local insurgents and tribal leaders, many of whom perceived the organisation as a lesser evil compared to the Iraqi army and security forces under Prime Minister Nouri al-Maliki.[64] The storming of Arab Sunni sit-ins across Central Iraq, the resulting Hawijah massacre in April 2013 and the clearing of Fallujah's year-long sit-in by the army in December 2013, have all strengthened that perception and facilitated tactical alliance with ISIS.

A third and final operational-level phase was a classic 'consolidation of power' via brutal liquidation of 'frenemies'. Most of the tribal leaders and insurgent commanders who assisted ISIS from January to March 2014 were targeted or liquidated afterwards by a mix of assassinations, IED-intensive tactics (including house-borne improvised explosive devices or HBIEDs), mass-executions, or imprisonment.

For example, armed militias close to the Iraqi Islamic Party (IIP) including Hamas of Iraq and local policemen close to the IIP were all selectively targeted.[65] In May 2014, the Fallujah Military Council publicly complained that ISIS was disarming and harassing its fighters as well as non-affiliated fighters inside Fallujah.[66] By the end of May 2014, ISIS dominated the six aforementioned organisations. The leaders of the organisations were in prison, had fled the town or been disarmed and marginalised. The most notable of these leaders were Sheikh Abdullah al-Janabi, who commanded the *Mujahidin* (Holy Fighters) *Shura* (Consultative) Council in the first battle of Fallujah in summer of 2004 and Brigadier-General Abu al-Walid, a former Baathist General who was a senior commander in the Fallujah Military Council.[67]

On the defensive: endurance and implosion

Combined Iraqi regular and irregular forces kicked off a major counteroffensive to take back Fallujah on 22 May 2016. At this point IS still had about 1,000 fighters in Fallujah, which had been under siege since February of that year. The Iraqi government and the Iraqi Shiite religious establishment managed to mobilise over 15,000 soldiers and fighters to liberate the town. The soldiers belonged to units from the first, the eighth, the tenth and the seventeenth divisions of the Iraqi Army as well as Iraqi Special Operations Forces (commonly known as the 'Golden Division', directed by the Iraqi Counter Terrorism Services or CTS), Iraqi Federal Police, the Popular Mobilisation Forces alliance of militias,[68] the Iranian Revolutionary Guard's al-Quds Force and strategic, intelligence and air-support from the American-led Combined Joint Task Force–Operation Inherent Resolve (CJTF-OIR). An estimated 29 CJTF-OIR aircrafts, including B-52s, A-10s, F-15s, F/A-18s, British Typhoons and UCAVs such as the MQ-1 Predator and MQ-9 Reaper pummelled IS positions, dropping more than 70 bombs per week. The battle to liberate Fallujah lasted till the end of June 2016.

Between June 2014 and June 2016, IS had built up defences in Fallujah, including preparations for tunnel-intensive, IED-intensive, sniper-intensive and GF-intensive tactics. IS relied heavily on various types of suicide bombings, including SVs, SVBIEDS and SMBIEDS. In one week in May 2016, IS launched 14 suicide operations.[69] However, the defeat in Fallujah was only a matter of time given the imbalance of power. Massively outnumbered and outgunned, the organisation attempted to retreat from Fallujah in a large convoy of over 200 armed vehicles and technicals. Given that IS has no complex air-defences and therefore could not cover the retreat; that decision was disastrous for the organisation. IS armed vehicles and technicals clustered in a traffic jam south of the town on 29 June 2016. 'There was no missing it', said Major-General Jay Silveria, deputy commander of US Air Forces Central Command.[70] At least 348 IS fighters were killed and about 200 IS vehicles were destroyed. This marked the end; the 2016 battle of Fallujah was decisively won by the Iraqi army and its allies. IS lost the town but did not stop its attacks there. As late as June 2019 – three years after the defeat – IS released a propaganda video showing night-time raids conducted by its fighters that resulted in gunpoint execu-tions of Iraqi army officers and *Sahwat* leaders in Fallujah's Karma district.[71]

The battlefront of Mosul: June 2014–July 2017

The first Battle of Mosul raged between 4 and 10 June 2014 and ended with a decisive ISIS victory. Even when compared to Fallujah and Ramadi, the Mosul battle was one of the most spectac-ular military upsets in the history of modern guerrilla warfare. ISIS was on the attack while outnumbered, with a manpower ratio that reached in the highest estimates 75-to-1 and in the lowest 10-to-1.[72] The organisation was also massively outgunned, attacking mainly with soft-skinned technicals, light weapons and IED-intensive tac-tics (including SVBIEDs and SVs). Moreover, the Mosul attack did not come as a surprise to the regular forces. According to a parliamentary investigation, Baghdad had prior intelligence about a forthcoming attack against the city in early May 2014.[73] Locally, both police and military intelligence services in Nineveh expected a forthcoming attack in June 2014. In addition, ISIS military com-mand was decapitated at the very beginning of the battle. Iraqi Federal Police units[74] successfully cornered Adnan al-Bilawi (Abu Abdul Rahman al-Bilawi), the head of ISIS Military Council, and almost arrested him. Al-Bilawi blew up himself using an SV. He is accused by the intelligence apparatus of the Iraqi Federal Police of masterminding both Nineveh and Anbar ISIS military campaigns.

Moreover, al-Bilawi was not even planning to take over the whole of Mosul. According to an IS documentary published on 10 June 2015 and entitled 'A Year Since the Conquest [of Mosul]', the initial plan was to take over two of the western districts (municipalities) of Mosul only. Out of the eight districts of Mosul, ISIS was only planning on occupying al-Rabee and New Mosul. The rest were occupied as a result of operational shifts during the battle, innovative tactics, the tenacity and brutality of ISIS fighters and a mix of corruption, perplexment and disarray amongst the regular forces.

An almost miraculous offensive

Similar to Fallujah, ISIS had used the SCCLC *modus operandi* earlier in Mosul. Initially, the organisation deployed about twenty small platoon-sized guerrilla formations to occupy two of the western districts of Mosul. Hence, most of ISIS' man- and firepower was concentrated in the west of Mosul, while diversionary attacks and disruptive forces were used in the northeast. On 5 June, a day after the neutralisation of al-Bilawi, ISIS fighters shelled the al-Zuhoor neighbourhood in northeast Mosul as a tactical diversion. By about midnight of 6 June, three convoys totalling less than 100 unarmoured pickup trucks and technicals, each carrying four-to-five fighters, advanced from al-Jazeera desert to the northwest of Mosul.[75] ISIS units rapidly advanced into the targeted Al-Rabee district (municipality) as planned, mainly in 17 Tammuz, al-Yarmuk and al-Nahrawan neighbourhoods. Almost simultaneously, the two intelligence HQs in the northeastern neighbourhood of al-Qahira were attacked in an attempt to 'blind' the regular Iraqi forces. Later on, this wrought havoc within command-and-control. Communications between Iraqi forces steadily broke down. Nineveh Operations Command had different battalions operating almost independently of one another in both eastern and western Mosul. At this point, ISIS overstretched its small units in the east, in a shift in the operational plan with the aim of reaching the eastern banks of Tigris River. The Iraqi regular forces fought back.

On 7 June, the Iraqi forces managed to regain some momentum; almost succeeded in cordoning off the northwest – where ISIS units were concentrated – with the help of air assaults. The air assaults, however, had a classic LAG effect.[76] They were not only inaccurate, but also accidently killed tens of local residents.[77] As a resident of western Mosul who witnessed the events said:

> The reaction on the ground was immediate. Al-Maliki's government in Baghdad was already unpopular and its army and federal police units in Mosul were perceived as sectarian and corrupt. Now, the army is killing the locals from the skies with American helicopters. What do you think will happen?[78]

In reaction to the Iraqi forces' counteroffensive, ISIS shut down the electricity plant in al-Yarmuk neighbourhood to control the damage. For much needed reinforcements, about 100 armed vehicles and technicals crossed over from ISIS-controlled territories in eastern Syria on 8 June. Under heavy Iraqi artillery and air-assaults, ISIS managed to advance into al-Uraybi neighbourhood. In the afternoon, an SLVBIED driven by a Saudi foreign fighter, Abu Omar al-Jazrawi, struck the Mosul Hotel, which was used as an operational headquarters in West Mosul.[79] The blow was significant. Several Federal Police commanders from the Fourth Brigade were killed in the explosion and it directly led to collective desertions within Federal Police units in central and southwestern neighbourhoods of the New Mosul district. Following this, the defences in western Mosul crumbled. Army soldiers and police officers were left without any plan or rallying point and fled when ISIS fighters began closing on their positions.

By 9 June, the nearby 17 Tammuz and al-Rabi neighbourhoods had fallen to ISIS, together with gradually almost all of western and south-western Mosul. Using SVBIEDs, the organisation's fighters stormed the Third Division base in al-Thawra neighbourhood from all four sides after a three-day siege. On the same morning, ISIS units fought their way to get closer to the Old City municipality. By the end of the day, ISIS had occupied the Mosul governorate building, the HQ of the Nineveh Police, the Mosul Airport and the one-mile-away al-Ghazlani Military Base. ISIS also overcame resistance in the east as many of the soldiers fled, believing their commanders to have abandoned them.[80] By 10 June 2014, ISIS was on its way to an unintended victory in Mosul, without the initial planning of its own leaders.[81] The organisation began an intense operation to secure the southern suburbs and disrupt a potential counteroffensive. Worried about airstrikes, ISIS stormed and seized both the Qayyarah Military Airbase and the Shirqat Airport. By the end of 10 June 2014, ISIS had taken over Mosul.

IS on the defensive

On 16 October 2016, the Iraqi government of Prime Minister Haider al-Abadi announced a military operation to liberate Mosul. It was dubbed 'Qadimun ya Ninawa' (We are Coming, Nineveh). On 9 July 2017, the Iraqi Prime Minister arrived in Mosul to announce the victory over IS. But IS continued to fight in the Old City district until August 2017. The organisation eventually lost Mosul after three years and two months. Given the conventional determinants of battle outcomes, including airpower, firepower, manpower ratios, as well as the familiarity of the allied forces with IS' innovative tactics (and hence the demise of the element of surprise), the size of the city and its lack of significant IS popular support;[82] the organisation had almost no chance of a strategic victory to endure in Mosul.

IS faced three critical challenges in Mosul: the airpower of the liberating forces, their overwhelming manpower and the declining impact of IS 'tactical innovations'. In terms of airpower, the coalition dropped 39,577 bombs in 2017.[83] During the battle of Mosul, 'coalition aircraft dropped more than 500 precision-guided munitions a week – even hitting as high as 605 weapons in one week', according to Brigadier General Matthew Isler, the deputy commanding general for air in the Combined Joint Forces-Land Component Command of Operation Inherent Resolve (OIR).[84] In the midst of the Mosul battle, military aircraft released more than 7,000 weapons against IS positions in January and February 2017 — the most of any two month period since OIR began in August 2014.[85] In terms of manpower, the ratio remained overwhelming against IS. The organisation's fighters were outnumbered, at a minimum estimate of 18-to-1 and at a maximum estimate of 38-to-1 in different neighbourhoods of the city.[86] Given that Mosul is over seven times larger than Fallujah and about five times the size of Ramadi, there was a remarkably low space-to-fighter ratio, reinforcing the factors against IS and determining the battle outcome.

Tactically, three categories of tactical innovation were significant in breaking momentum and slowing down the advance of the Iraqi regular forces and supporting militias, costing them thousands of deaths during the ten-month battle. These were the IED-intensive tactics (especially SVBIEDs), the GFs/SGFs-intensive tactics and sniping. IS also introduced UCAV/drone-intensive tactics in Mosul as of October 2016. Despite the relatively limited damage they caused, the drone-borne IEDs (DBIEDs) and drone-intensive tactics nonetheless had an impact on the Iraqi troops' morale.[87]

In terms of IED-intensive tactics, IS relied heavily on the usage of SVBIEDs. According to IS' *A'maq* Agency, the organisation carried out 482 SVBIED attacks in the whole battle of Mosul between 19 October 2016 and 12 July 2017.[88] The average of about three SVBIEDs every two days does not explain the whole picture, though. IS began the first month of fighting (17 October to 17 November) with 124 SVBIED attacks, a sustained four SVBIEDs per day for 30 days. This is the highest recorded rate of SVBIEDs ever used in Iraq and elsewhere in the world.[89] By the last month of fighting[90] (18 June to 17 July 2017), *al-Naba'* newsletter of IS only reported 11 SVBIEDs (about one every three days). The SVBIEDs were consistently dropping despite spikes in the fifth (80 SVBIEDs), sixth (32 SVBIEDs) and seventh (44 SVBIEDs) months of fighting.[91] Despite the havoc wreaked by these IED-intensive tactics, many of the VBIEDs and SVBIEDs tactics were effectively countered by new tactics and weapons of the regular Iraqi forces. These include dispersing the advancing regular forces, forwarding early-warning/anti-SVBIEDs screening forces, enhancing coordination between joint terminal attack controllers (JTACs) on the ground and the coalition air forces to early-strike the SVBIEDs before they reach their targets. In terms of GFs/SGFs-intensive tactics and sniping, IS had built up multiple combat branches specialised in urban small-unit warfare.[92] These combat units – operating as GFs or SGFs – displayed a relatively competent degree of mobility, manoeuvrability and lethality, compared to the Iraqi forces who were relatively slower and unable to quickly react to the urban warfare challenges in Mosul's streets. IS units combined defensive IED emplacements, with tunnel complexes, and light and medium mortars. Overall, during the ten months of fighting, *al-Naba'* newsletter claimed over 110 operations conducted by GFs or SGFs as well as over 1,629 sniping operations.[93]

The Ramadi battlefront: May 2015–February 2016

ISIS launched simultaneous attacks aiming to control Ramadi and Fallujah as early as January 2014, even before the establishment of IS in July of that year. As the capital of the Anbar Province and a critical juncture on the road to Baghdad, Ramadi was both strategically and symbolically important for ISIS. In January 2014, the organisation managed to seize several municipal buildings and police headquarters in various districts of the town. However, unlike in Fallujah, it failed to capture the whole of Ramadi. Another round of the Ramadi battle ensued in October 2014, after the initiation of Operation Inherent Resolve (OIR). On 16 October 2014, *Wilayat al-Anbar* (IS Anbar Province) published a series of photos showing IS units patrolling Ramadi's neighbourhoods, while other reports estimated that about 60 per cent of Ramadi's districts, including southern, western and northern ones, were either contested or controlled by IS.[94]

Between October 2014 and May 2015, IS proceeded according to the SCCLC *modus operandi* it had used in Fallujah and in Mosul (softening and creeping, attempted coalition-building and then liquidating-consolidating).[95] However, the second phase specifically was much harder to implement successfully in Ramadi, given the history of the town with IS predecessors. Hence, ISIS/IS failed in its efforts at 'coalition-building'. This was one of the reasons the town withstood attacks after January 2014.[96] Numerous clans and tribes of Ramadi had historical vendettas with both IS and its predecessors, some of which went back to 2005 and earlier.[97]

The 'softening and creeping' attacks escalated in October 2014. IS launched a sustained S/VBIED campaign featuring over 67 attacks in the next few months. By late November, IS units were about 160 yards away from the centre of Ramadi, but Iraqi regular forces and supportive tribal militias were able to repel them and successfully defend the centre. IS managed to mobilise fighters from other towns in neighbouring Anbar and Salahuddin 'Provinces' and from eastern Syria via the Jazeera desert to renew the assault on Ramadi in early May.[98] The decisive battle for Ramadi occurred between 14 and 17 May 2015. On the night of 14 May, IS launched its attack on the centre of the town from all four directions, using at least 13 armoured SVBIEDs and SLVBIEDs, and 12 formations of GFs and SGFs (*inghimasiyyun*).[99] Although the Iraqi army and anti-IS militias have endured stronger attacks involving over 20 armoured SVBIEDs, IS's tenacious war of attrition after January 2014 took its toll on them.[100] On 15 and 16 May 2015, IS continued the attack relentlessly with a focus on the main Government Complex and Anbar Operations Command (AOC), south and southeast of the Euphrates.

Aside from the common tactics earlier used in Fallujah and Mosul, IS also benefited from employing tactics involving armoured front-end loaders (AFELs). In other battles, IS has used AFELs to

punch holes in fortifications such as concrete blast barriers or earth-
berms, allowing both armoured and/or soft-skinned SLVBIEDs
to pass through before striking and destroying targets. But in the
Ramadi town-centre battle in May 2015, IS converted the AFEL
into an armoured SVBIED to attack an Iraqi army HQ based in an
abandoned building in the town centre.[101] The eight-floor building was surrounded by two belts of concrete blast barriers in order
to stave off any attacking SVBIEDs. For speed, the suicide bomber
in the AFEL-turned-SVBIED knocked over and moved aside both
layers of concrete barriers and then detonated the AFEL inside the
building. In the three days of fighting, IS repeated that tactic with
slight modifications – at least four times – in areas and locations controlled by the Iraqi regular forces and allied local militias. By 17 May
2015, the regular forces had had enough and started withdrawing.
On the same day, the spokesperson of the governor of Anbar stated
that Ramadi had fallen to IS.

On the defensive: the costly liberation of Ramadi

The US-led Coalition and the Iraqi forces escalated the campaign to
liberate Ramadi beginning from July 2015. About six months later,
Ramadi was liberated from IS but with a heavy cost: the destruction
of about eighty per cent of the town. One of the reasons for the
wide-scale destruction was that IS had constructed a strong defensive
posture during the six months of its occupation. Still, the organisation was massively outnumbered by an estimated manpower ratio
that ranged from 16-to-1 to 25-to-1.[102]

Tactically and operationally, IS constructed a multi-layered defence
based on IEDs, booby-traps, interlocking sniping positions and networks of underground and inter-building tunnels. The tunnels permitted the GFs and the SGFs to rapidly change positions, undetected
by the Iraqi soldiers or the surveillance UCAVs. IS also used IEDs as
defensive minefields. It littered the main road in Ramadi and the surrounding areas with IEDs and then aimed machine guns and mortars at
the field of explosives to protect them from the demining and disposal
efforts of the Coalition.[103]

By February, the same critical variables that IS faced in Mosul had
their impact in Ramadi: the airpower of the liberating forces, their
overwhelming manpower and the declining impact of IS' 'tactical
innovations'. The airpower was a major component in defeating IS

in Ramadi, to the extent that the OIR spokesperson said that it constituted 80 per cent of the reasons why the city was recaptured from
IS.[104] The airstrikes permitted the Iraqi Special Operations Forces[105]
and allied militias to advance with other Iraqi army units playing a
limited role.[106] Overall, the coalition conducted over 800 airstrikes in
Ramadi from July 2015 to late February 2016.[107] By then, IS tactical
innovations were partly understood; thus the element of surprise had
little-to-no impact. For example, many of the protected IED-fields
were destroyed via a new counter-tactic: an explosive rope attached
to a rocket to blast a lane through the explosive field, therefore
allowing the regular forces to advance rapidly in that lane.[108] Also,
tens of the VBIEDs, SVBIEDs, AFELs and their likes were destroyed
by coalition airstrikes, as listed in the OIR's strike releases.[109]

Overall, from the above overview of the three battlefronts in
Fallujah, Mosul and Ramadi, some clear fighting patterns emerge.
At a macro-level, ISIS/IS rose up and pulled off multiple military
upsets in Iraq due to a successful military build-up that was facilitated by a repressive socio-political environment and counterinsur-
gency/counterterrorism mishaps. At the meso- and micro-levels,
the organisation's capacity to innovate tactically and to operation-
ally-strategically shift (between terrorism, guerrilla and conventional
ways of warfare) allowed it to attain almost miraculous operational
victories, when it was both outnumbered and outgunned, and with
neither an airforce nor adequate air-defence systems. A strategic
victory, however, was unattainable, given the overwhelming manpower, firepower, airpower and growing familiarity with IS' tactical
innovations.

How ISIS/IS fights in Iraq: battlefronts analysis

As outlined above, ISI/ISIS/IS had pulled off military upsets in
Iraq, using surprisingly innovative ways of fighting. The mixture
of creative tactics and strategic shifts (from terrorism to guerrilla to
conventional warfare and back) resulted in surprising operational
victories in the short- and mid-term. The following section analyses some of the tactical and operational patterns employed by ISIS/
IS in Iraq, partly based on its own publications and statements.
The analysis is based on surveying the details of armed operations
conducted by ISIS/IS in Fallujah, Mosul and Ramadi between late
2013 and late 2019.

Fallujah: tactical shifts

The analysis of ISIS/IS media reporting on the Fallujah battlefronts shows that the organisation relied mainly on thirteen categories of tactics to fight the regular forces and anti-ISIS militias in town. During the Fallujah battle of December 2013 and January 2014, ISIS did not report consistently on its tactics via *al-Naba'* newsletter and other media outlets/releases. This was partly due to the 'coalition-building' operation that the organisation implemented at this phase. Hence, the Global Terrorism Database (GTD) reported over 115 unclaimed attacks during and after the January 2014 battle.[110] During the Fallujah January 2014 battle, at least six guerrilla formations belonged to ISIS, out of 18 others fighting regular forces in and around the town. Between 30 December 2013 and 4 January 2014, ISIS did not use IEDs as a main battlefield weapon. Its fighters relied more on tactical alliances with other armed organisations[111] and on attacking the regular Iraqi forces and police by GFs, artillery, technicals and anti-aircraft autocannons used mainly against ground targets. By one estimate, ISIS formations used light and medium mortars at least eight times and anti-aircraft autocannons at least five times during the battle.[112]

The reported categories of tactics have significantly changed during the 2016 battle of Fallujah, as demonstrated by the sample listed in Table 2.2.[113]

Table 2.2 IS Reported Categories of Tactics in Fallujah in 2016

Category of Tactics	Reported Usage
SVBIEDs	45
Sniping	23
Unguided Rockets	17
ATGMs	16
MANPADS	13
Artillery	13
HBIEDs	11
Guerrilla Formations (GFs)	8
Suicidal Guerrilla Formations (SGFs)	8
Anti-aircraft Autocannons	5
IEDs	4
VBIEDs	2
Close-quarter Assassinations	1
Total	**166**

By 29 December 2016, *al-Naba'* and other IS media outlets had reported 45 uses of SVBIED-led or SVBIED-intensive tactics, 23 of sniping-intensive tactics, 17 of unguided rocket attacks, 16 of ATGM-intensive tactics and 13 operations using MANPADs in the Fallujah battle. The tactical and operational changes were clear. IS shifted from relying on short-term tactical alliances with Sunni armed organisations, to relying on its own SVBIEDs in 27 per cent of the attacks reported by its media. The second category of tactics involved sniping. It was used in 14 per cent of IS reported operations in the 2016 Battle of Fallujah. This was followed by tactics involving unguided rockets and ATGMs (10 per cent each), and then MANPADs and artillery (8 per cent each). In Fallujah, the organisation has employed different methods of fighting in 2016, compared to 2014. Relatively, it fought distinctively compared to Mosul and Ramadi battlefronts.

Mosul: tactical innovations and strategic shifts

As described above, the battle of Mosul provided unique insights into how IS has developed its ways of fighting in dense urban terrain between the June 2014 and October 2016–July 2017 battles of Mosul. Prior to June 2014, ISIS used its traditional mix of urban terrorism and guerrilla warfare in 'softening and creeping' operations to lower the morale of the Iraqi regular forces in Mosul.[114] In IS jargon, 'softening and creeping' operations were paving the way for *tamkin* (consolidation/domination).[115] As illustrated in its *al-Naba'* article on guerrilla warfare, IS aimed to inflict damage, lower morale, destroy communications and supply lines between the army and the federal police units in the different districts of Mosul and surrounding areas.[116] That way each unit would be responsible for its own security and too occupied to support other units. The environment became too dangerous to move forces between Mosul's districts because of the chance of GFs/SGFs ambushes.[117] As a part of the 'soften and creep'/ SCCLC *modus operandi* – just before the battle of Mosul on 4 June 2014 – IS has conducted at least six SVBIED-intensive attacks, ten assassinations of high-ranking officers, nine attacks by GFs and two attacks by SGFs in a mix of terrorism and guerrilla warfare tactics.

During the battle to remain in Mosul, IS has mainly relied on fifteen categories of tactics in the ten-month battle as illustrated below in Table 2.3 (overleaf).[118]

Table 2.3 IS Reported Categories of Tactics in Mosul, October 2016 to July 2017

Category of Tactics	Reported Usage
Sniping	1,629
SVBIEDs	482
Anti-aircraft Autocannons	112
Suicidal Guerrilla Formations (SGFs)	110
ATGMs	71
Guerrilla Formations (GFs)	44
Tunnel-intensive	41
HBIEDs	34
IEDs	33
Unguided Rockets	32
Artillery	25
MANPADS	23
Close-quarter Assassinations	21
VBIEDs	8
Total	**2,665**
Drones	+200

The above-mentioned attacks reported by *al-Naba'* newsletter did not include drone-intensive tactics or direct drone attacks. However, IS has staged over 200 attacks using various types of drones, documented by other open sources and IS propaganda videos.[119] Still, the most reported battlefield tactic used prolifically by IS in the battle of Mosul (2016–17) was sniping, followed by SVBIEDs (either SVBIED-intensive or SVBIED-led tactics). Indeed, the battle of Mosul featured the most prolific usage of SVBIEDs as a battlefield weapon in history.[120] As detailed above, in the first 100 days of the Mosul offensive, IS had employed over 270 SVBIEDs in an attempt to slow down the liberating forces and break their momentum with as much damage as possible. The initial attacks were quite devastating for the Iraqi regular force, including their most COIN-competent elite-force, the CTS 'Golden Division'.

But the Iraqi forces showed rapid adaptability through fortifying positions by using earth berms and makeshift roadblocks.[121] They also destroyed parts of all five of the Tigris bridges to interdict the flow of SVBIEDs from western to eastern Mosul and positioned M1 Abrams tanks at intersections of roads. That positioning helped in destroying many SVBIEDs by the main gun rounds of the tanks. The Iraqi regular forces were increasingly reliant on a pre-emptive anti-SVBIED tactic: air-striking to crater roads where the forces suspect an SVBIEDs potential route. IS units countered some of these measures by using more than one SVBIED in each attack. The first one or ones would breach the defensive berm or roadblock, and the rest would aim for the target(s). IS also used camera-equipped unmanned non-combat aerial vehicles (non-combat drones) to bypass roadblocks and guide the SVBIEDs onto targets, using live-video and -audio feed. Despite the prolific use of SVBIEDs, IS did not extensively deploy static IEDs and dense minefields in urban Mosul. One possible explanation for this is the significant presence of civilians in Mosul, compared to the depopulated Ramadi.

In addition to the staggering numbers of SVBIEDs used, the battle of Mosul has also witnessed another 'first': the most intensive use of UCAVs in a battlefront employed by any ANSA in the world. According to one account, IS had flown over 300 drone missions in one month during the battle for Mosul.[122] General Raymond Thomas, the Head of the US Special Operations Command, explained that an adaptive IS 'enjoyed tactical superiority in the airspace under the [American] conventional air superiority'. American conventional air superiority has stopped any attack on American ground forces from enemy aircraft since 15 April 1953 during the Korean War.[123] IS changed that by using commercially available drones, flying under 3,500 feet. The US Air Force do not own the airspace at such a low-level. IS media outlets did not provide comprehensive metrics on drones' activities during the battles of Mosul. But, in the midst of the Mosul Battle, during the first 20 days of February 2017, IS media outlets released 113 images of IS drone-attacks.[124] In one day in December 2016, IS employed 70 drones almost simultaneously, right under American conventional air superiority, bringing the Iraqi troops' liberation efforts to a temporary halt.[125]

IS units have mainly used a mix of quadcopters and fixed-wing drones available off-the-shelf. The main four munition types that were usually dropped or fired from IS drones were IEDs, 40mm

munitions, hand-grenades and RPG warheads.[126] IS also used drones
as decoy Ieds. For example, during the battle of Mosul, an IS drone
appeared to have crash-landed near a Peshmerga position. When the
Peshmerga fighters took it apart, the battery pack was an IED in
disguise. It detonated, killing two of the Peshmerga fighters.[127] But
IS' combat drones are not as lethal when compared to other categories of tactics such SVBIEDs, other IED-intensive tactics, SGFs and
sniping. Hence, IS used drones as tactical decoys in combinations
with SVBIEDs and SGFs. For example, drone strikes were used to
distract Iraqi soldiers on the ground from the greater danger of an
approaching SVBIED, which detonates causing much more damage.[128] In Mosul, the overall number of successful IS drone attacks
was relatively reduced when the United States deployed counter-
drone jamming systems up to the front lines.[129]

Ramadi: tunnels and tenacity

Despite the liberation of Ramadi – or what was left of it – in February
2016, IS launched sustained attacks throughout the year. Table 2.4
shows a sample of categories of tactics used by IS in Ramadi during
2016.[130] Seventy-seven of the reported attacks were launched between
1 March and 29 December 2016, after the liberation of the town and
the victory of the Iraqi regular forces and their allies. The majority of
the seventy-seven post-liberation attacks were conducted with either
IEDs or ATGMs. However, there was one complex attack in March
2016 that was executed with four SGFs, two GFs and three SVBIEDs.

   The way that IS fought in Ramadi, both offensively and defensively, was also slightly nuanced compared to elsewhere. Offensively,
the OIR had already started and the anti-IS coalition was relatively
familiar with the earlier tactics employed by the organisation. The
tenacious attrition warfare of IS in May 2015 and earlier proved to be
effective, particularly the sustained attacks by SFGs, GFs and sniper
formations. These tactics were used defensively as well, in combination with up-armoured SVBIEDs and technicals (with mounted
12.7mm heavy machine guns as well as 14.5mm and 23mm anti-aircraft autocannons) as shown in two propaganda videos on the battles
of Ramadi entitled 'Azm al-Kumat 1 and 'Azm al-Kumat 2 (Resolve
of the Protectors 1 and Resolve of the Protectors 2). IS used over
30 up-armored SVBIEDs in the offensive battle to occupy Ramadi
in May 2015. Some of these SVBIEDs were used to breach the Iraqi

Table 2.4 IS Reported Categories of Tactics in Ramadi in 2016

Category of Tactics	Reported Usage
IED-intensive	26
SVBIED-intensive	17
Guerrilla Formations (GFs)	15
HBIEDs	14
Artillery	14
Unguided Rockets	13
ATGMs	11
Suicidal Guerrilla Formations (SGFs)	11
Sniping	6
Close-quarter Assassinations	6
VBIED-intensive	3
Total	**136**

defenses, along with the conventional infantry breaching sequence of suppress, obscure, secure, reduce and assault (SOSRA). IS was combining terrorism tactics with conventional tactics similar to those the US Marine Corps employed in the 2015 Ramadi battle.[131]

Defensively, over 80 per cent of Ramadi was depopulated by the time the Iraqi forces and their allies had begun their liberation efforts. As a result, IS was able to litter the town with hundreds of static, covert IEDs. The sample of 26 IED-intensive operations reported in *al-Naba'* newsletter (issues 12 to 61) do not adequately reflect the scale and the intensity of the static IED usage in Ramadi in 2016. Sources in the Iraqi army estimated clearing more than 150 IEDs in the town after its liberation. OIR sources reported that thousands of IEDs and booby-traps were planted by IS in the town.[132]

Al-Naba' newsletter did not also report on the scale and the intensity of tunnel warfare employed by IS in Ramadi. Offensively, IS fighters dug a network of over 20 tunnels – both connected and independent – in their efforts to occupy the town before May 2015 and afterwards. These include short (82 yards) tunnels, such as the one dug from al-Huz neighbourhood to the Police Command HQ in the heavily guarded government compound, as well as long ones.[133] The long ones in Ramadi exceeded 1,650 yards, such as the tunnel dug under the house of Hamid al-Hais, the head of the Anbar Salvation Council.[134] The house was used as an operational HQ for coordinating between Iraqi regular forces and the Sunni tribal militias fighting IS in Ramadi.[135] IS planted tens of IEDs estimated to weigh over a tonne and blew-up the house.

In addition to planting or delivering IEDs, the tunnels in Ramadi were also used to deliver SGFs, GFs and various weapon systems. Most of the Ramadi tunnels had reinforced walls within them and some were equipped with electricity and ventilation systems. Hence, defensively, it allowed IS fighters to hide from airstrikes, move without detection from overhead surveillance and withdraw from close-quarters combat when necessary. Overall, tunnel warfare and its labyrinths offered key tactical advantages to IS fighters entrenched in Ramadi, both defensively and offensively. The tunnels posed a challenge for the Iraqi regular forces, which lacked the knowhow to detect and map the underground labyrinth, making small-unit operations even more complicated.

The future of IS insurgency in Iraq

In March 2019, the Trump administration announced the defeat of IS after a five-year campaign featuring some of the most intense urban battles since World War Two. The declaration was too optimistic, more suitable for domestic consumption in an electoral campaign than for reflecting the realities on the ground in Iraq and elsewhere. Both of the American Defense Intelligence Agency (DIA) and the United Nations Security Council (UNSC) estimated in mid-2018 that IS still retains up to 30,000 fighters in Iraq and Syria only.[136] In 2019, the international anti-IS coalition estimated that the organisation has approximately 11,000 fighters in Iraq.[137] And even after the liberation of Fallujah, Mosul, Ramadi and other towns and villages, the Coalition has conducted 13,331 strikes in Iraq from August 2017 to mid-2019.[138]

To compare, ISI – a predecessor of IS – was left with a maximum of 1,000 fighters in 2010, when it was considered almost destroyed. It still made a major comeback featuring the takeover of Fallujah, Mosul, Ramadi and other Iraqi towns and villages in about four years. In 2019, assuming the accuracy of the coalition's estimate, IS was not just about 11 times stronger in terms of manpower in Iraq, but it is also a much more experienced organisation in terms of strategic shifts, tactical innovations, operational arts of warfare and local and foreign fighters' recruitment. IS recruitment of foreign fighters was estimated to be at the rate of 50 per month in 2017–18,[139] compared to an average of five foreign fighters per month in 200–7 for ISI. Overall, IS endured as a military organisation still capable of executing sustained urban terrorism and guerrilla warfare operations.

Despite that, IS in Iraq has certainly been degraded by the OIR, the Iraqi regular and irregular forces and their allies. From January 2018 to May 2019, IS took responsibility for only 28 VBIED- and MBIED-intensive attacks, an average of about two per month. These terror operations covered Iraq, from Nineveh in the north to Babel in the south and included attacks in Baghdad. This is a significant downgrade from an average of two SVBIEDs per day in 2016, and an average of almost three SVBIEDs per day during the 'Breaking the Walls' campaign (July 2012 to July 2013). By March 2020, a year after President Trump's aforementioned declaration, IS had claimed responsibility for 1,146 attacks in seven out the 19 Iraqi governorates; a sustained overall-average of three armed operations per day for 365 days.[140] This represented both an overall increase and a concerning indicator.

The degradation of IS was costly in blood and treasure. The Mosul battle of 2016–17 was described as the 'most significant urban battle since World War Two' by Lieutenant-General Stephen Townsend, the commander of the CJTF-OIR.[141] The Iraqi Counter-Terrorism Service's 'Golden Division' sustained casualty rates of up to 60 per cent in the Battle for Mosul of 2016–2017.[142] All fourteen battalion commanders of the Golden Division were killed in action against IS. The Iraqi Kurdish Peshmerga suffered at least 2,000 casualties and the Iraqi regular forces suffered over 10,000 causalities during the campaign against IS in Iraq.[143] By March 2018, after the liberation of Mosul, OIR had cost the US $23.5 billion.[144]

In terms of the future, IS and its predecessors in Iraq has shown a previous capacity to rebuild their military capabilities after being on the brink of collapse. The organisation strategically shifts between conventional, guerrilla and terrorism ways of warfare to avoid annihilation. From 2006 onwards, the organisation was able to operationally and tactically innovate to pull-off multiple upsets against stronger foes. These innovations include modified, upgraded and original urban terrorism, guerrilla warfare and conventional military tactics. As of May 2017, IS shifted its strategy again to focus on guerrilla warfare and terrorism. That shift meant dissolving units of its existing structures and dispersing forces into rural and remote areas, including areas where IS had built subterranean tunnel complexes. IS' existing rural tunnel networks serve as weapons depots and foodstuffs, assisting the organisation in executing its guerilla war and terrorism strategies. [145] For example, on 18 December 2017, the Iraqi regular forces seized enough explosives for fifty VBIEDs in an underground tunnel in Mutaibija on the borders of Diyala and Salahuddin Governorates. [146]

In terms of shifting to guerrilla warfare, IS had launched a campaign entitled 'The Conquest of Attrition' on 31 May 2019. The organisation aimed to wear down the Iraqi forces to destroy their will to fight and temporarily, rather than permanently, hold territory. The campaign represents a modification of the SCCLC *modus operandi*. In terms of urban terrorism tactics, IS has targeted Fallujah, Mosul, Ramadi, as well as Kirkuk City and Tikrit with VBIEDs in 2019. The organisation still retains VBIED-making capacity in Iraq, indicating that its manufacturing networks were not completely destroyed by the OIR. On 2 June 2019, the 'Iraq Province' of IS released a video[147] showing a nighttime raid led by IS commander Abdul Azim al-Iraqi.[148] The raid targeted a meeting in the house of a tribal militia leader. Iraqi army officers were among the attendants. All the attendees were killed at gunpoint with silenced handguns and AK-47 assault rifles. The location was significant: al-Karma District in northeast of Fallujah. IS was still conducting terror raids in Fallujah, three years after its liberation. These attacks should not be construed as a mere sign of desperation, but as an operational level *modus operandi* that aims to punish/liquidate local militia leaders and Sunni loyalists to Baghdad, terrorise the population into submission and soften/wear down anti-IS forces in Iraq, whether regular or irregular.

Finally, on a macro-level, the future of IS in Iraq is likely to be affected by the urban destruction and uninhabitability of major parts of Mosul, Ramadi, Fallujah and other towns, mainly as a result of the battles of liberation. IS is more likely to capitalise on such an environment. The destruction of these urban centres may lead to further alienation and marginalisation of the (mainly Arab-Sunni) locals. Hence, there is a potential for IS to recruit and radicalise on a sectarian-Wahhabist basis.

Notes

1. Pat Work, 'Combating ISIS in Mosul', lecture given at the Royal Military Academy at Sandhurst, Berkshire on 8 July 2019.
2. John Ismay, Thomas Gibbons-Neff and C. J. Chivers, 'How ISIS Produced Its Cruel Arsenal on an Industrial Scale', *New York Times*, 10/12/2017, https://nyti.ms/2jzaIBo
3. As a reminder, the acronyms stand for: Islamic State in Iraq and Sham (ISIS); Islamic State in Iraq (ISI); Hilf al-Mutayyibin (HM – or Coalition of the Good); *Majlis Shura al-Mujahidin* (MSM – or Holy Fighters Consultative Council); al-Qaida in the Lands of the Two Rivers or al-Qaida in Iraq (AQI); and *al-Tawhid wa'l-Jihad* in Iraq (TJI – Monotheism and Struggle).
4. Arabic *kunya* (nickname) translated as 'the father of Mus'ab from Zarqa' (a city in northwest Jordan). His real name is Ahmed al-Khalayleh.
5. For excellent analyses of IS and its predecessors in Iraq see: Azmi Bishara, *The State Organisation (Acronymed ISIS): A General Framework and a Contribution to Help Understand the Phenomenon*, [in Arabic] vol. 1 (Doha: the Arab Centre for Research and Policy Studies, 2018); Azmi Bishara (editor), *The State Organisation (Acronymed ISIS): Foundation, Discourse and Practice*, [in Arabic] vol. 2 (Doha: the Arab Centre for Research and Policy Studies, 2018). Haider Said, 'The Way to the Fall of Mosul', [in Arabic] *Siyasat Arabiya*, no. 10 (September 2014); David Kilcullen, *Blood Year* (Oxford: Oxford University Press, 2016); Gareth Stansfield, 'Explaining the Aims, Rise, and Impact of the Islamic State in Iraq and Al-Sham', *Middle East Journal*, vol. 70, no. 1 (2016); William McCants, *The ISIS Apocalypse* (New York: St. Martins Press, 2015); Ahmed Hashim, 'From al-Qaida Affiliate to The Rise of The Islamic Caliphate: The Evolution of the Islamic State of Iraq and Syria (ISIS)', *Policy Report, RSIS*, 12 /2014. Accessed on 15/9/2016, at: https://bit.ly/32YzWet
6. The timeframe usually begins when ISIS/IS takes over a town/city. It ends when ISIS/IS loses it.
7. Nabih Bulos, 'Islamic State Has Been Cranking Out Car Bombs on an Industrial Scale For The Battle Of Mosul', *New York Times*, 25/2/2017, https://lat.ms/2SUn4Bt; Thomas Joscelyn, 'Analysis: Islamic State Claims Historically High Number of Suicide Attacks in 2016', *FDD's Long War Journal*, 3/1/2017, https://bit.ly/2OxuZWA
8. See 'Translation of Old Al-Zarqawi Interview, Says God's Law Must Rule 'Entire World', *Open Source Report*, 6/12/2006, *Open Source Center*. Quoted in: Ahmed Hashim, 'From al-Qaida Affiliate to The Rise of The Islamic Caliphate: The Evolution of the Islamic State of Iraq and Syria (ISIS)', *Policy Report, RSIS*, 12 /2014. Accessed on 15/9/2016, at: https://bit.ly/32YzWet. Complete Arabic version can be found in *Minbar al-Tawhid wa al-Jihad*, https://bit.ly/333EkJ1.

9. Zarqawi meant by that term the lack of state-sponsorship and therefore territories to retreat to (like the case of Pakistan during the *Mujahidin* insurgency in Afghanistan), and the lack of enough popular support to allow AQI to establish an effective military front and to retreat to these areas when necessary.

10. 'Interview with Abu Mus'ab al-Zarqawi', [Arabic] *Minbar al-Tawhid wa al-Jihad* (Hijri-year 1427 or 2006). Accessed on 15/4/2019, at: https://bit.ly/2xzgJFh

11. Ibid. In 2006, Zarqawi estimated that the 'war in Iraq' will be won by tactical and operational combinations of suicide bombers and guerrilla formations. It was an early indicator of the forthcoming military impact of suicide guerrilla formations (SGFs) or the *inghimassiyun*.

12. Brad Knickerbocker, 'Relentless Toll to US Troops of Roadside Bombs: The IED Has Caused over a Third of the 3,000 American GI Deaths in Iraq', *Christian Science Monitor*, 2/1/2007, p. 1; John Bokel, 'IEDs in Asymmetric Warfare', *Military Technology*, vol. 31, no. 10 (10/ 2007), p. 34.

13. *Mafariz* can be also translated as 'detachments'.

14. '*Waqafat 'ind Ahadith* – 5 (Pauses in Conversations – Part 5)', *Al-Naba'*, no. 101, 22/11/ 2017, p. 8.

15. Stephen Biddle, Jeffrey A. Friedman and Jacob N. Shapiro, 'Testing the Surge: Why Did Violence Decline in Iraq in 2007?', *International Security*. vol. 37, no. 1 (Summer/ 2012), pp.7–40,

16. Omar Ashour, 'Viewpoint: How Islamic State is Managing to Survive', *BBC News*, 14/12/2015, https://bbc.in/31aqK4G.

17. Hamed al-Zawi was a former Iraqi police officer from al-Zawiyah village, close to the town of Haditha in the Anbar Governorate.

18. *Waqafat 'ind Ahadith*, p. 8.

19. The real name is Abdul Mun'im al-Badawi, a former mid-ranker in the Egyptian al-Jihad Organisation.

20. *Waqafat 'ind Ahadith*, p. 8.

21. Ibid; Former Major-General in the Iraqi Army, Interview by the author, Istanbul, 7 February 2019.

22. In Arabic: *Masa'id al-Mughaffalin*.

23. Michael S. Schmidt, 'Suicide Bombs in Iraq Have Killed 12,000 Civilians, Study Says', *New York Times*, 2/11/2011, https://nyti.ms/2YyXG9G .

24. This period corresponds to the year 1434 in the *Hijri* Lunar calendar, which started in the middle of November 2012. See: New magazine from the Islamic State of Iraq and al-Sham: al-Bina' Magazine', *Jihadology*, 31/3/2014, https://bit.ly/316ulAW/; 'al-Furqan Media Presents a News Bulletin from Islamic State of Iraq and al-Sham: 'Harvest of Operations for the Year 1433 H in Iraq', *Jihadology*, 14/8/2013, https://bit.ly/2KmlNz0 .

25. Other monthly reports were compiled and found in the following issues of *al-Naba'*, nos. 50, 51, 52 and 53. See also: Abu Khabab al-Muhajir, '*Nashrat al-Naba'*, 2/2014, https://bit.ly/2OArONW; Abu Khabab al-Muhajir, '*Nashrat al-Naba'*, 1/2014, https://bit.ly/2K8YKca.

26. As mentioned in the first chapter, IS' definition of an 'armed operation' remains unclear. Some of the operations listed in their databases include more than one armed attack. For example, the organisation would list an operation involving a suicide bombing followed by mortars and sniping tactics as one single operation, as opposed to three attacks.

27. The classifications were done according to the data published by the older version of *al-Naba'* newsletter, issues numbered 50, 51, 52 and 53, respectively.

28. MBIEDs are motorcycle-borne IEDs.

29. SMBIEDs are suicide motorcycle-borne IEDs.

30. See IS Nineveh Province's video production entitled '*Wa Lanahdiyannahum Subulana* (And We Shall Surely Guide them to Our Ways) showing some of the manufacturing and modification of munitions and weapons by IS: 'New Video Message from the Islamic State: We Will Surely Guide Them to our Ways- Wilayat Ninawa', *Jihadology*, 17/5/2017, https://bit.ly/2ypnC90.

31. For details of the airstrikes and their targets, as well as the degradation and destruction inflicted on IS by OIR see: Operation Inherent Resolve, 'Targeted Operations to Defeat ISIS: Airstrike Updates', *U.S Department of Defence*, https://bit.ly/2Zqazji.

32. Ernest Barajas Jr., a former Marine explosive ordnance disposal technician who has worked with ordnance-clearing organisations in areas occupied by the Islamic State. Quoted in: John Ismay, Thomas Gibbons-Neff and C. J. Chivers, 'How ISIS Produced Its Cruel Arsenal on an Industrial Scale', *New York Times*, 10/12/2017, https://nyti.ms/2jzaIBo.

33. Most of these armoured pieces were captured form the Iraqi army in the summer of 2014. Some of the T-55s were captured from the Syrian army and Syrian armed opposition formations. Former Army of Islam Commander, Interview by author, Istanbul, 7 November 2016.

    Also see: 'Weapons of ISIS - Islamic State Infantry Weapons, Vehicles and Artillery', *Military Factory*, https://bit.ly/3381myH; Mehmet Kemal Firik, 'ISIS's Weapon Inventory Grows', *Daily Sabah Mideast*, 2/7/2014, https://bit.ly/2ysc6d5

34. The M198 has a maximum range of 18 miles and a maximum rate of fire of four rounds per minute. It is uncommon among guerrilla arsenals.

35. The towed Type 58 gun has a maximum a range of 17 miles. It is also uncommon to find it in guerrilla arsenals.

36. A man-portable air defence surface-to-air system (MANPADS) that can be a serious threat to low-flying aircraft. FIM-92 Stingers require specialised maintenance and care, however. Apparently, IS did not have the necessary resources.

37. An infrared-homing, surface-to-air missile from the Russian-made *Igla* family. The US Department of Defence's designation for it is SA-16; the Russian designation is 9K310 Igla-1E.

38. An infrared-homing, surface-to-air missile from the Russian-made *Igla* family as well. SA-18 is the US Department of Defence's designation for

it. The Russian designation is 9K38 Igla missile. It is more accurate, with a
longer range compared to the SA-16.

39. Abu Rajab, former FSA Fighter who fought ISIS/IS in Idlib, Aleppo and
    Raqqa. Interview by the author, Istanbul, 9 October 2016.

40. An infrared-homing, surface-to-air missile from the Russian-made *Strela*
    family. SA-7 is the US Department of Defence's designation for it. The
    Russian designation for it is 9K32 *Strela-2*. It is an older generation of
    Soviet MANPADS, compared to the *Iglas*.

41. 'Weapons of the Islamic State: A Three-Year Investigation in Iraq and
    Syria', *Conflict Armament Research* (12/2017): pp. 183–4.

42. Only two of these were documented by the Conflict Armament Research
    (CAR). Ibid., p. 184.

43. Such as the defunct *Jaysh al-Khilafah* (the Caliphate Army), *Jaysh al-Istishadiyyin*
    (Martyrdom-ists Army or 'Suicide-Fighters' Army), and *Jaysh Dabiq* (Army of
    Dabiq). In 2020, the only (smaller) regiment-sized formation that survived is
    referred to as *Jaysh Khalid* (Khalid's Army) and operates mainly in Iraq with
    some units crossing over to operate in eastern Syria.

44. Within ISI/ISIS/IS, a sector or a *qati'* is a military-geographical unit within
    a province, which can range from a neighbourhood in a town to a group of
    towns and their environs.

45. The statement rhymes in colloquial Iraqi-Arabic.

46. Thom Shanker, 'Al-Qaeda Leaders in Iraq Neutralised, US Says', *The New
    York Times*, 4/6/2010. Accessed on 15/10/2016, at: https://nyti.ms/2XphSd4

47. Qassim Abdul-Zahra and Lara Jakes, 'Al Qaeda Iraq Strength Musters',
    Associated Press, 10/10/2012. Accessed on 16/9/2016, at: https://bit.
    ly/2Kh7pYX.

48. The military council was at some point renamed *Hay'at al-Harb* (War
    Department).

49. Both Tarkhan Batirashvili (a Georgian-Kist ISIS/IS commander known as
    Abu Omar al-Shishani) and Gulmurod Khalimov (a Tajik IS commander
    known as Abu Omar al-Tajiki) have allegedly held the position of 'War
    Minister', according to the Pentagon. However, during the restructuring of
    the ISI in the summer of 2011 by Haji Bakr, the positions of 'War Minis-
    ter' and 'Chief of Staff' were abolished. Their reinstatements have not been
    confirmed by ISIS/IS publications, except by 'informal' documents of IS
    insiders alleging that al-Shishani was the 'Emir of the War Department'.
    Al-Shishani, a former sergeant in the 'Special Reconnaissance Group' of the
    Georgian Army, was killed in July 2016. Russian military officials alleged to
    have killed al-Tajiki in April 2017. This was not confirmed by the Pentagon
    back then. Most of the Russian official statement on the killings of IS and
    insurgent commanders have proven to be inaccurate, unless confirmed by
    other source(s). See: The 'informal' document on IS' *Qanat Arshif al-Siyar*
    (Biographies Archival Channel), https://bit.ly/2Yz4d49

50. See ISI's documents on breaking walls found here: https://bit.ly/336DyLe
    Also see: Jessica Lewis, 'Al Qaeda in Iraq is Resurgent', *Middle East Security
    Report*, no. 14 (9/2013), https://bit.ly/1G7kP0L, p. 8.

51. Tazoult-Lambese prison-break in 1994 in Batna during the Algerian civil war is numerically comparable, given the large number of escapees (about 1,000 prisoners). However, the Lambese operation was based on deception and recruitment of guards as opposed to sheer force. Based on the testimonies of the escapees, the large number of freed prisoners was a direct result of a fluke and – as opposed to ISI – the GIA failed to organise or even to recruit the overwhelming majority of the escapees; many of whom were criminal convicts unaffiliated with the insurgency.

52. '824 Prisoners Broken Out By Force since Al-Baghdadi Announced "Breaking the Walls" Campaign',]In Arabic[Al-Hayat, 25/7/2013. Accessed on 17/9/2018, at https://bit.ly/2GFEfSl

53. See: Wael Issam, 'ISIS Gains Control of Fallujah after Detaining Military Council Officers',]In Arabic[Al-Quds Al-Arabi, 28/6/2014. Accessed on 15/9/2016, at https://bit.ly/2YAw8Ay. Former Major-General in the Iraqi Army, Interview by the author, Istanbul, 7 February 2019.

54. Ibid.

55. Ibid.

56. Bill Roggio, 'ISIS Parades on Outskirts of Baghdad', The Long War Journal, 1/4/2014. Accessed 1/6/2018 at: https://bit.ly/2Zkwzfw

57. The organisation faced significant local resistance between January and March 2014, even after it declared Fallujah to be part of its territory during Friday prayers on 3 January 2014.

58. See the sections below on the battlefronts of Mosul and Ramadi. Also see the chapters on the Syrian and the Libyan battlefronts.

59. Most notably the southern neighbourhoods and southern rural outskirts of Fallujah and Euphrates River communities south of Fallujah, such as al-N'imiya and 'Amiriy areas. AQI had some support in these areas in 2006.

60. Anti-ISIS elements included anti-AQI elements, Sahwat figures, non-cooperative and less-cooperative tribal figures and policemen.

61. On Hawijah see: Tim Arango, 'Dozens Killed Across Iraq as Sunnis Escalate Protests Against Government', The New York Times, 24/4/2013. Accessed on 18/9/2016, at https://nyti.ms/34sd0p5.

62. Kamal Namaa, 'Iraqi Militants Kill at Least 18 Soldiers, Including Commander', Reuters, 21/12/2013. Accessed on 18/9/2016, at https://reut.rs/31ejSDH

63. 'Anbar Governorate: 85 Bridges were Destroyed by ISIS and We Need 160 Million Dinars to Rebuild Them', Al Mada press, 25/10/2015. Accessed on 18/9/2016, at https://goo.gl/SY4yRw

64. International Crisis Group Iraq, 'Falluja's Faustian Bargain', Middle East Report, no. 150, 28/4/2014. Accessed on 15/11/2018, at https://bit.ly/333SUQR

65. Former Major-General in the Iraqi Army, Interview by the author, Istanbul, 7 February 2019.

66. Wael Issam, 'ISIS Gains Control of Fallujah after Capturing Military Council Officers',]In Arabic[Al-Quds Al-Arabi, 28/6/2014. Accessed on 15/12/2016, at https://bit.ly/2YAw8Ay

67. Ibid.

68. A broad coalition of mainly Shiite militias which have been effective in fighting ISIS/IS in Iraq. However, some of these militias have been accused of committing major human rights violations, as well as sectarian and criminal practices.

69. Wilayat al-Fallujah, '14 *'Amaliya Istishhadiyya* [14 Martyrdom Operations]', *Al-Naba'*, no. 36, 24/5/2016, p. 6.

70. Oriana Pawlyk , 'Diverting to Fallujah from Syrian Town was Right Call to Target ISIS, General Says', *Air Force Times*, 15/7/2016. Accessed on 15/11/2017, at https://bit.ly/2KdNtWU
    For more details on that operation see: Christiaan Triebert, 'An Open Source Analysis of the Fallujah 'Convoy Massacre'(s)', *Bellingcat*, 6/7/2016. Accessed on 15/10/2019, at https://bit.ly/29nqLup

71. 'New Video Message from the Islamic State: We will Surely Guide Them to Our Ways- *Wilayat Niwaya'*.

72. At the high-end of the estimate, these manpower ratios include the Third Division of the Iraqi Federal Police in Mosul and its support units, which were estimated to be 30,000-strong. This is in addition to the Second Division of the Iraqi army and its support units which were estimated to be another 30,000-strong. Both divisions operate under Nineveh Province Operations Command. ISIS initial attacking force was in the range of 400–600 fighters. At a later stage of the battle, ISIS cells inside Mosul joined them, and on 8 June reinforcements of about 400–500 fighters arrived from eastern Syria.
    At the low-end of the estimate, Lieutenant-General Mahdi Al-Gharrawi, the former Commander of the Iraqi Federal Police in Nineveh, claimed in a *Reuters* report that there were only 10,000 soldiers 'in reality' in all combat divisions in Mosul. See: Ned Parker, Isabel Coles, Raheem Salman, 'Special Report: How Mosul fell - An Iraqi general disputes Baghdad's Story', *Reuters*, 14/10/2014. Accessed on 15/9/2016, at https://reut.rs/2GIXieC

73. 'The investigation regarding the Fall of Mosul reveals the responsibility of Maliki and Other High Ranking officials',]In Arabic[*Al Watan News*, 19/8/2015. Accessed on 11/9/2016, at https://bit.ly/2LUn03P

74. This was a joint operation between the Intelligence Apparatus of the Federal Police and the Rapid Response Force of the Iraqi Ministry of Interior (Police and Security Ministry).

75. 'Announcement of ISIS in Mosul', [in Arabic] *Al Jazeera*, 12/6/2015. Accessed on 12/2/2018 at: https://bit.ly/2LTLhXR

76. For the LAG and its impact in COIN operations, see the section entitled 'Why Weaker Insurgents Survive or Beat Stronger Incumbents?' in the first chapter.

77. 'How Did Extremists Take Over One of Iraq's Biggest Cities in Just Five Days?' *Niqash*, 10/6/2014. Accessed on 15/10/2016, at https://bit.ly/2YF18iQ

78. A Witness from al-Rabee neighbourhood in western Mosul, Interview by the author, 8 January 2019.

79. 'This is how ISIS Controlled al-Mosul', [in Arabic], *Akhbar al-Yawm*, 12/6/2015. Accessed on 10/9/2016, at: https://bit.ly/2K8oBkz ; 'Documents of the Second Man Answers: How ISIS Controlled al-Mosul', *Qasioun*, 5/6/2018. Accessed on 11/9/2018, at: https://bit.ly/2Zqxviv

80. Berguen Sfanson, 'Interesting Facts about the fall of Mosul to ISIS a Year Ago',]In Arabic[Trans. Rim Nejmi, *Deutsche Welle*, 1/6/2015. Accessed on 14/9/2016 at: https://bit.ly/2KbDBy6

81. According to an IS documentary released on 10 June 2015 and entitled 'A Year Since the Conquest', the initial plan was to take over the western districts (municipalities) of Mosul only. Out of the eight Mosul districts, IS was planning to occupy only two: al-Rabee and New Mosul. For more details on the initial plans of ISIS, based on the documents confiscated from al-Bilawi's hideout, see: 'Documents of the Second Man Answers: How ISIS Controlled al-Mosul'.

82. The Iraqi Institute of Administration and Civil Society Studies conducted a small opinion-poll in the city of Mosul. It found that in June 2014 the percentage of those who believe that IS represents their views or interests did not exceed 10 per cent. See: Munqith Dagher, 'Combating Daesh: We are Losing the Battle for Hearts'. [In Arabic], *IIACSS Reports* (Iraq: IIACSS, December 2015). See also: Arab Opinion Index survey on ISIL, in: 'The 2017–2018 Arab Opinion Index: Main Results in Brief', *Arab Centre for Research and Policy Studies*, at: https://bit.ly/38xi9OF, pp. 34–6.

83. Since the OIR operations began in Iraq and Syria on 8 August 2014, the Coalition's Air-Force has flown more than half of the total of 167,912 sorties during the Mosul battle.

84. Stephen Losey, 'With 500 Bombs a Week, Mosul Airstrikes Mark "the Most Kinetic" Phase of ISIS Air War So Far', *Airforce Times*, 28/3/2017. Accessed on 15/7/2017, at: https://bit.ly/2MCwQXR; Stephen Losey, 'Airstrikes Against ISIS Hit All-time High', *Airforce Times*, 13/9/20017. Accessed on 15/7/2017, at: https://bit.ly/2x7urwE

85. Ibid.

86. Estimates are based on the open source materials published during the ten-month battle in Mosul and its suburbs. See for example: Tim Hume, 'Battle for Mosul: How ISIS Is Fighting to Keep its Iraqi Stronghold', *CNN*, 25/10/2016. Accessed on 15/7/2017, at: https://cnn.it/2ZwnNeq; Isabel Coles, John Walcott, Maher Chmaytelli, 'Islamic State Leader Baghdadi Abandons Mosul Fight to Field Commanders, US and Iraqi Sources Say', *Reuters*, 8/3/2017. Accessed on 15/7/2017, at: https://reut.rs/2ZuPeFD; '*Irak : L'opération pour Reprendre Mossoul des Mains de l'EI est Lancée*', *Le Monde*, 17/10/2016. Accessed on 15/7/2017, at: https://bit.ly/31cW2bf.

87. Don Rassler, 'The Islamic State and Drones: Supply, Scale and Future Threats', *Combatting Terrorism Center at West Point* (July 2018), at: https://bit.ly/2O8QUiP

88. This figure includes the involvement of some of the 110 SGFs listed, as documented by *al-Naba'* newsletter issues number 78, 79, 80 and 81. Like other figures in the database, this is a minimum estimated sample. According to one senior CENTCOM officer, IS in Mosul has detonated over 600 SVBIEDs and SLVBIEDs between October and December 2016 during the battle of Mosul. This estimate suggests a higher figure. See for example: Mark Perry, 'How Iraq's Army Could Beat ISIS in Mosul, But Lose the Country', *Politico*, 15/12/2016. Accessed on 16/7/2017, at https://politi. co/38r6v65

89. The only comparable rate would be found in Syria. For more details, see the following chapter.

90. Excluding the IS pockets which was besieged in the Old City district and lasted till August 2017.

91. Hugo Kaaman, 'Islamic State Statistics on its SVBIED Use from Late 2015 Through 2017, Including the Battle of Mosul', *Hugo Kaaman Open Source on SVBIEDs*, 18/8/2018. Accessed on 15/8/2019, at https://bit. ly/2YDYoSP

92. Issa Smisim, '*Al-Inghimasiyun*: the Striking Force of Jihadist Organizations',]In Arabic[*al-Arabi al-Jadid* 22/12/2014. Accessed on 15/7/2017, at: https://bit.ly/2vkqPoO; '*Al-Inghimassiyun* ISIS's Most Lethal Weapon',]In Arabic[*al-Khalij Online*, 17/8/2014. Accessed on 15/7/2017, at: https:// bit.ly/2T27FPs

93. Compiled by the author and his research team. The range of operations are listed in *al-Naba'* newsletters from issue number 61 to issue number 112.

94. Bill Roggio and Caleb Weiss, 'Islamic State Photos Highlight Group's Grip on Ramadi', *FDD's Long War Journal* (16/10/ 2014). Accessed on 15/7/2017, at: https://bit.ly/2LY0EOT

95. Ghassan Al-Issawi (Spokesperson of Anti-IS Tribal Coalition in Ramadi). Interview by Qusai Shafiq, *Al-Ahd* Channel, Iraq, 27/8/2017. Accessed on 15/11/2018, at https://bit.ly/2MKOHvR; General Rashid Al-Falih (Commander of the Popular Mobilization Forces in al-Anbar). Interview by Qusai Shafiq, *Al-Ahd* Channel, Iraq, 27/8/2017. Accessed on 15/9/2017, at: https://bit.ly/2MKOHvR

96. Ibid.

97. Al-Issawi, Interview. Also see: Carter Malkasian, 'Anbar's Illusions: The Failure of Iraq's Success Story', *Foreign Affairs*, 24/6/2017. Accessed on 15/7/2017, at: https://fam.ag/2MD9r8y; Khalid al-Ansary and Ali Adeeb, 'Most Tribes in Anbar Agree to Unite Against Insurgents', *The New York Times*, 19/09/2018. Accessed on 15/7/2019, at: https://nyti. ms/2H9n6AE

98. Dalf Al-Kubaisi, an official in al-Anbar Governorate, estimated in an interview that 500 technicals arrived from eastern Syria and central Iraq to Anbar between late April and early May 2015. See: Manaf al-Abidi, 'The Story of the Fall of Ramadi',]In Arabic[*al-Sharq al-Awsat*, 26/6/2015. Accessed on 15/7/2017, at: https://bit.ly/2yC1BUC

99. Former Major-General in the Iraqi Army, Interview by the author, Istanbul, 7 February 2019; Ghassan Al-Issawi (Spokesperson of Anti-IS Tribal Coalition in Ramadi). Interview by Qusai Shafiq on *Al-Ahd* Channel, Iraq, 27 August 2017. Hesham al-Hashimi (Former Iraqi government security advisor). Interview by Qusai Shafiq, *Al-Ahd Channel*, Iraq, 27/8/2017, at: https://bit.ly/2MKOHvR. Also see: Patrick Martin, Genevieve Casagrande, Jessica Lewis McFate, 'ISIS Captures Ramadi', *Institute For The Study Of War* (18/5/2015). Accessed on 15/7/2019, at: https://bit.ly/2YoM3Tc

100. al-Abidi, 'The Story of the Fall of Ramadi', [In Arabic].

101. 'al-Furqan Media presents a New Video Message from the Islamic State: "And They Gave Zakah"', *Jihadology*, 17/6/2015. Accessed on 15/7/2017, at: https://bit.ly/2T3NvEF .

102. The initial estimate was about 1,200 IS fighters in October. By, the last week there was about 250. See: Falih Hassan et al., 'Celebrating Victory Over ISIS, Iraqi Leader Looks to Next Battles', *The New York Times* 30/12/2015. Accessed on 15/11/2017, at: https://nyti.ms/2Zf7kdz Raed al-Hamed, 'The Battle of Ramadi: The Coalition's Gains and ISIS' Tactics',]In Arabic[*Aljazeera Centre for Studies*, 13/2/2016. Accessed on 15/11/2017, at: https://bit.ly/2KhyPzj

103. Colonel Steve Warren, a Baghdad-based spokesman for the US Department of Defense, quoted in Paul Blake, 'Ramadi Assault: How a Small Change in Tactics helped Iraqi forces', *BBC*, 22/12/2015. Accessed on 15/7/2017, at: https://bbc.in/2YK3tW0

104. Nancy A. Youssef and Shane Harris, 'How ISIS Actually Lost Ramadi', *Daily Beast*, 30/12/2015. Accessed on 15/7/2017, at https://bit.ly/2MCUALw

105. They are also known as 'Counter Terrorism Forces' (operating under the CTS) and the 'Golden Division'.

106. Youssef and Harris, 'How ISIS Actually Lost Ramadi'.

107. See the OIR airstrike updates here: Operation Inherent Resolve, 'Targeted Operations to defeat ISIS: Airstrike Updates', *U.S Department of Defence*, https://bit.ly/2Zqazji. Also see: Youssef and Harris, 'How ISIS Actually Lost Ramadi'.

108. Warren quoted in Blake, 'Ramadi Assault: How a Small Change in Tactics helped Iraqi forces'.

109. Operation Inherent Resolve: Strike Release, *U.S Department of Defence*, https://bit.ly/335e3u5

110. See the Global Terrorism Database (GTD) on Iraq for the year 2014: https://bit.ly/2Yq6yz9 .

111. For a list of these organisations, see above in the section entitled: 'The Battlefront of Fallujah: January 2014 and June 2016'.

112. Former Major-General in the Iraqi Army, Interview by the author, Istanbul, 7 February 2019.

113. This dataset is a part of the 'Islamic State Ways of Warfare' Database (ISWD-Fallujah14).

114. Some of these guerrilla warfare tactics are described in detail in four arti-
     cles published by *al-Naba'* newsletter of IS, issues number 179, 180, 181
     and 182.
          Also see: Lt. Col. Craig Whiteside and Vera Mironova, 'Adaptation and
     Innovation with an Urban Twist Changes to Suicide Tactics in the Battle for
     Mosul', *Military Review*, 11/12/ 2017. Accessed on 15/8/2018, at: https://
     bit.ly/338G3wB
115. See: 'Temporarily Conquering Cities: *Modus Operandi* for Holy Fighters',
     [in Arabic] *Al-Naba'*, no. 180, 3/5/2019, p. 9.
116. Ibid., p. 9.
117. Ibid., p.9.
118. This dataset is a part of the 'Islamic State Ways of Warfare' Database
     (ISWD-Mosul1617). Some of the categories of tactics are gleaned from
     al-Naba' newsletter issues numbered from 62 to 112 respectively.
119. On the drone campaign, see: Don Rassler, 'The Islamic State and Drones:
     Supply, Scale and Future Threats', *Combatting Terrorism Center at West
     Point*, 7/ 2018, https://bit.ly/2O8QUiP
          See Also: Ben Sullivan, 'The Islamic State Is Pioneering a New Type of
     Drone Warfare', *Vice*, 2/2/2017. Accessed on 16/7/2017, at: https://bit.
     ly/2yAmsTy; Don Rassler, Muhammad Al-`Ubaydi and Vera Mironova,
     'The Islamic State's Drone Documents: Management, Acquisitions, and
     DIY Tradecraft', *Combatting Terrorism Centre at West Point*, 31/1/2017.
     Accessed on 15/7/2019, at: https://bit.ly/2KaYqus; Ben Sullivan, 'The
     Islamic State Conducted Hundreds of Drone Strikes in Less Than a Month',
     Vice, 21/2/2017, https://bit.ly/2YBAyDr ;'New video message from the
     Islamic State: We Will Surely Guide Them to Our Ways - *Wilayat Niwaya'*.
120. Hugo Kamaan, 'Islamic State SVBIED Development since 2014', paper
     presented at the annual conference of the Strategic Studies Unit enti-
     tled 'Militias and Armies: Developments of Combat Capacities of Armed
     Non-State and State Actors', Arab Centre for Research and Policy Studies
     (ACRPS), Doha, 24 February 2020, https://bit.ly/2XqDk1j
121. Some of these were made up of abandoned civilian vehicles or even parts
     of destroyed VBIEDs and T-walls.
122. Not all of these drones were directly used in combat. Some of them were
     used for surveillance, propaganda, or for guiding SVBIEDs.
123. Peter Grier, 'April 15, 1953', *Airforce Magazine* (June 2011), p. 54.
124. Ben Sullivan, 'The Islamic State Conducted Hundreds of Drone Strikes in
     Less Than a Month', https://bit.ly/1oteczo
125. David B. Larter, 'SOCOM Commander: Armed ISIS Drones were
     2016's 'Most Daunting Problem', *Defense News*, 16/5/2017. Accessed on
     11/8/2019, at: https://bit.ly/2KoaYg4
126. Former Peshmerga Commander, Interview by the author, Prague,
     February 2017. Also see: Nick Waters, 'Types of Islamic State Drone
     Bombs and Where to Find Them', *Bellingcat*, 24/5/2017. Accessed on
     11/8/2019, at: https://bit.ly/2qpmjp1

127. Michael Schmidt and Eric Schmitt, 'Pentagon Confronts a New Threat From ISIS: Exploding Drones', *The New York Times*, 12/10/2016. Accessed on 11/8/2019, at: https://nyti.ms/2Z8fg0b

128. Former Peshmerga Commander, Interview by the author, Prague, February 2017.

129. Ibid.

130. This dataset is a part of the 'Islamic State Ways of Warfare' Database (ISWD-Ramadi16).

131. IS units did not obscure using the common smoke-grenades tactics in this battle. However, they used diversionary attacks to obscure.

132. Colonel Steve Warren, 'Department of Defense Press Briefing by Col. Warren via Teleconference from Baghdad, Iraq', *US Department of Defence*, 29/12/2015. Accessed on 11/8/2019, at: https://bit.ly/2MzO5ZG; Susan Jones, 'Ramadi liberated, But Booby Trapped', *CNS News*, 11/2/2016. Accessed on 12/8/2019, at: https://bit.ly/2GIgg56; 'Iraq: British Contractor Killed Clearing Mines In Ramadi', *Sky News*, 22/8/2016. Accessed on 12/8/2019, at: https://bit.ly/2GJXJVX

133. '*al-Aan* News Reports on ISIS's Use of Tunnels to Infiltrate the City of Ramadi in Anbar', *al-Aan* News, al-Aan TV, 26/2/2015. Accessed on 11/8/2019, at: https://bit.ly/2yBGdyI

134. One of the armed Arab-Sunni tribal coalitions fighting IS in al-Anbar.

135. Omar Janabi, 'Tunnels of Death. . .The New Weapon of ISIS', [Arabic] *al-Khaleej Online*, 16/03/2015. Accessed on 11/9/2017, at: https://bit.ly/2Z5YAXi

136. Although, the CJTF-OIR assessed that IS is down to 1,000 fighters in December 2017. See: 'Less Than 1,000 IS Fighters Remain in Iraq and Syria, Coalition Says', *Reuters*, 27/12/2017. Accessed on 11/8/2019, at: https://bbc.in/31fAL0A; See also the United Nations Security Council's (UNSC) estimate in: https://undocs.org/S/2018/705. Office of the Inspector General, 'Overseas Contingency Operations: Operation Inherent Resolve and Operation Pacific Eagle- Philippines Report to the United States Congress: 1 April 2018–30 June 2018', *US Department of Defense*, 6/8/2018. Accessed on 14/8/2019, at: https://bit.ly/2GIXCK4.

137. See the estimate in the United Nations Security Council's (UNSC) tenth report of the Secretary-General on the threat posed by ISIL (Da'esh) dated 4 February 2020 in: https://undocs.org/S/2020/53.

138. See: Operation Inherent Resolve-Combined Joint Task, *US Defence Department*, https://bit.ly/31iDarE.

139. Office of the Inspector General, 'Operation Inherent Resolve and Other Overseas Contingency Operations: October 1, 2018 - December 31, 2018', *US Department of Defense*, 4/2/2019. Accessed on 11/8/2019, at: https://bit.ly/2YDG4Ft, p. 21; 'Foreign Fighters Continue to Join ISIS in Syria, US Joint Chiefs Chair Says', *Defense Post*, 16/10/ 2018. Accessed on 12/8/2019, at: https://bit.ly/2Mx3vhr

140. The targeted governorates are Anbar, Nineveh, Kirkuk, Salahuddin, Diyala, Baghdad, and Babel.
141. Department of Defense Briefing by General Townsend via Telephone from Baghdad, Iraq', *US Department of Defense* (28/3/2017). Accessed on 13/8/2019, at: https://bit.ly/2yz1hFY; Jonathan Marcus, 'Mosul: Have Combat Changes Increased Civilian Casualties?' *BBC*, 29/3/2017. Accessed on 14/8/2019, at: https://bbc.in/2YiIUV0
142. David M. Witty, 'Iraq's Post-2014 Counter-Terrorism Service', *Washington Institute for Near East Policy* (October 2018).
143. 'Press Conference by Special Presidential Envoy McGurk in Erbil, Iraq', *US Department of State*, 4/9/ 2017. Accessed on 14/8/2019, at: https://www.state.govpress-conference-by-special-presidential-envoy-mcgurk-in-erbil-iraq
144. 'Cost of War Through 31/3/2018', *Federation of American Scientists*, https://bit.ly/2LCsPhX
145. Derek Henry Flood, 'From Caliphate to Caves: The Islamic State's Asymmetric War in Northern Iraq', *CTC Sentinel* (September 2018). Accessed on 14/8/2019, at: https://bit.ly/2LZzRlA
146. 'Explosives Found in Mutaibijah Enough to Booby-Trap 50 Cars' *Al-Sumaria TV*, 18/12/2017.
147. It was the first IS video from Fallujah since April 2017.
148. According to the video, the raid was conducted on 12 November 2018; See: 'New Video Message from the Islamic State: "Then They Will Be Overcome – Wilayat al-Iraq, Al-Fallujah"', *Jihadology*, 2/6/2019. Accessed on 15/8/2019, at: https://bit.ly/2yD04O0

3

Explodes and Expands: How the 'Islamic State' Fights in Syria

Raqqa is very important for them [ISIS]. They [ISIS fighters] won't let it go.

> Abu Hamza, Eastern Region Commander of
> *Ahrar al-Sham*, January 2014[1]

The use of IEDs and explosives is more dense [in Raqqa] than other cities . . . including the Iraqi cities of Mosul, Ramadi and Fallujah.

> Colonel Ryan Dillon, Spokesman of Operation Inherent
> Resolve (OIR), 17 October 2017[2]

In five months, they [Marines] fired 35,000 artillery rounds on ISIS targets in Raqqa, more than any other Marine or Army battalion since the Vietnam war.

> Army Sergeant Major John W. Troxell, Senior Adviser to the
> Chairman of the Joint Chiefs of Staff, US Army,
> 6 February 2018[3]

Enhanced old tactics in troubled new context: the 'leap' from Iraq to Syria

On 23 January 2012, *Jabhat al-Nusra li Ahl al-Sham* (JN), or 'The Support Front to the People of the Levant' was publicly declared in a dramatic propaganda video-statement.[4] Its leader *al-Fatih* (The Conqueror) Abu Muhammad al-Julani (Ahmad Hussein Ali al-Shar')[5] outlined the JN's aims and methods within the Syrian armed revolution against the ruling regime of Bashar al-Assad. The JN propaganda video left out a few important details. The first is that the organisation was a Syria-based front for the Islamic State in Iraq (ISI).[6] It was established on the orders of the ISI leader, Abu Bakr al-Baghdadi and the commander of his Military Council at the time, Haji Bakr. Another important detail is related to al-Julani, the pronounced leader of the JN. The man was the former 'governor' (*wali*) of Nineveh Province in ISI, known by the *nom de guerre* 'Aws al-Mawsili'.[7] He is not from Mosul (as al-Mawsili connotes). His ancestors hailed from the village of *fiq* in the Golan Heights (hence the title al-Julani). His family was internally displaced in Damascus after the 1967 war and the Israeli occupation of the Golan Heights.

Al-Julani crossed the Iraqi–Syrian borders (from the Nineveh governorate to al-Hasaka governorate) on a late night in August 2011, five months after the largely unarmed Syrian revolution started against the Assad regime.[8] Al-Julani was accompanied by five other ISI commanders, all wearing suicide vests and armed with light weapons. The two Syrians were Salih Hamah (not to be confused with Saleh al-Hamawi)[9] and Anas Khattab. The two Jordanians and the Palestinian commanders were Iyad Tubasi (Abu Julaybib),[10] Mustafa al-Salih (Abu Anas al-Sahaba)[11] and Abu Omar al-Falastini, respectively. Maysara al-Juburi (Abu Mariya al-Qahtani), an experienced Iraqi ISI commander, was already based in Deir Ezzor since 2009. Contrary to many reports, he was not among the founding squad coming from Iraq.[12] He did join them later on.

The initial strategic objective of the ISI squad was to capture as much territory and resources as possible in Syria, then use them for multiple objectives among which was to launch a war of attrition in Iraq. This plan was not new in jihadism per se. In the 1990s, the Libyan Islamic Fighting Group (LIFG) had a similar plan during the Algerian civil war. The LIFG referred to it as *al-wathba* (the leap).[13] The LIFG's 'leap' was disastrous in terms of planning and execution. ISI's 'leap', however, was not. Between August 2011 and April 2013, the tiny squad of ISI commanders created two of the most effective military organisations on the Syrian insurgent scene, impacting the armed conflicts in Syria, Iraq and elsewhere.

The JN, with ISI assistance, rapidly developed its combat capacities from relying on static IEDs and S/VBIEDs used primarily in urban terrorism operations to mastering a mix of guerrilla/revolutionary warfare and conventional tactics.[14] On 8 April 2013, an audio statement by the then ISI leader, al-Baghdadi, was released. The latter finally declared that the JN was a part of his organisation and that he was merging both groups under the newly formed 'Islamic State in Iraq and Sham' (ISIS).[15] Al-Baghdadi's declaration ignited a chain of rapid reactions, including a quick rejection of the merger by al-Julani on 9 April 2013,[16] a declaration of allegiance to al-Qaida (AQ) in the same audio-statement of al-Julani,[17] the defection of the majority of JN fighters to ISIS, a failed intervention by AQ's leadership to contain the rivalry, and then brutal hostilities between ISIS fighters and the remaining JN loyalists throughout 2013 and 2014.[18] On 5 February 2014, AQ's General Command declared that it had no organisational ties with ISIS and that it was not responsible for any of its actions. On 23 February 2014, ISIS assassinated an alleged 'envoy' of AQ in Syria, Muhammad Bahayya (Abu Khalid al-Suri), in a complex operation involving five ISIS fighters wearing suicide vests.[19]

The chaotic war between ISIS/IS and JN did not prevent these two organisations from becoming main military actors in the Syrian civil war, with a combined control or area-denial of over 50 per cent of Syria's territories in 2015.[20] These successors of ISI in Syria (JN, ISIS and IS) were able to endure and expand their geographical scope, tactical military capacity, operational intensity and durations, regional scale, quality of propaganda and communications between 2012 and up to 2020. The expansion and the endurance happened despite the relatively limited popular support for both organisations[21] and the lack of state-sponsorship, compared to the Russian and Iranian support for the regime's forces and the Turkish, Saudi and American support for different brands of opposition forces. While it is true that the Assad regime pursued a policy of selective and intermittent 'collusion and collision' with ISI, ISIS and IS, before and during the war in Syria; this policy alone, by no stretch of imagination, explains the military rise of the organisations.

This chapter provides an overview of the birth and the military build-up of ISIS/IS in Syria, as of 9 April 2013. It aims to explain how ISIS/IS was able to gradually develop their combat capacities and the resulting combat and military effectiveness in Syria. The chapter then focuses on describing and analysing the battlefronts of Raqqa Governorate between 2013 and 2019. Raqqa City, the governorate's provincial capital, was the first 'capital' of the organisation. Arguably, IS had shown its maximum combat capacities in Syria during its occupation of the governorate and in defence of its 'capital'. In any case, the battlefronts of Raqqa had shown high-levels of IS combat performance and a wide-range of IS tactics. Indeed, when units in the Syrian armed opposition decided to remove or destroy ISIS in Raqqa on 2 January 2014, they received a warning. 'If the battle is

not well-planned and if victory is not guaranteed, do not fight them
(ISIS units). Raqqa is very important for them. They [ISIS fighters]
won't let it go',[22] warned Abu Hamza, the then Commander of *Ahrar
al-Sham* in the Eastern Region (al-Hasaka and Deir Ezzor).[23]

   This chapter is partly based on interviews with Syrian rebels and
soldiers who fought against IS in eight Syrian governorates: Raqqa,
Deir Ezzor, Aleppo, Homs, Hama, Latakia, Damascus and Rif
Dimashq. It is also based on documents, audio-visual and photographic releases produced by ISIS/IS in Syria. The majority of these
releases were produced by three out of the (maximum of) ten ISIS/IS
Syria-based 'provinces', proclaimed between 2014 and 2018.[24] The
three are *Wilayat al-Raqqa* (Raqqa Province), *Wilayat al-Khayr* (The
Good Province/Deir Ezzor) and *Wilayat Halab* (Aleppo Province).
The chapter also relies on official documents released by the US
government, Operation Inherent Resolve (OIR) and other open-
source materials.

   The chapter is composed of five sections. The following section
gives an overview of the military build-up of ISIS/IS in Syria since
its establishment in April 2013. The third section outlines the details
of the battlefronts of Raqqa Governorates within specific timeframes.
The fourth section analyses how IS fights in Syria, using data and
observations from the Raqqa battlefronts as well as others, such as
Deir Ezzor and Aleppo Governorates. Finally, the concluding section reflects on the future of IS insurgency in Syria after losing Raqqa
and other territories and shifting back to mainly guerrilla and terrorism tactics.

Fashioning force: an overview of IS military build-up in Syria

The IS military build-up in Syria was impressive, even when compared to the build-up in Iraq where the organisation was almost
destroyed by 2010 and then rose up to militarily occupy the second-
largest city in 2014. The trajectory of fashioning force in Syria was
different, however. As opposed to the micro- and meso-level security
policies in Iraq (and in Libya),[25] the policies of the Assad regime in
Syria between 2003 and 2011 selectively facilitated the existence of,
and provided assistance to, logistical support networks of ISI and its
predecessors as well as to other anti-American armed organisations in
Iraq.[26] In different timeframes, the Assad regime had also selectively

cracked down on some of these networks and arrested many of their affiliates. Still, ISI had an already-existing network and safehouses before 2011. This network had limited capacities compared to the organisation's network in Iraq.

That changed rapidly due to a five-pillar, operational-level *modus operandi* of building-up and fashioning a military force. For simplification, it will be referred to as iALLTR as it will be repeated in other countries. Between 2011 and 2013, JN/ISI in Syria attempted (successfully and otherwise) to:

1) Gather intelligence[27] (i): Given that a significant percentage of the JN-ISI/ISIS fighters were non-locals,[28] and given that the organisation's origins and founding squad came from Iraq, it had to build intelligence capacities to map out targeted territories and local organisations.
2) Absorb/Recruit (A/R): absorb already-existing, like-minded organisations and factions operating in Syria, and populate their units by both individual and collective recruitment of youth;
3) Loot (L): loot parts of both Assad regime's and armed opposition's arsenals and modify, convert and upgrade them;
4) Lead (L): lead this intelligence, absorption, recruitment, looting and other operations by relying on the experiences of ISI commanders as well as others who joined from combat zones abroad;
5) Transfer (T): transfer knowhows and disseminate knowledge of tactical innovations, shifting ways of warfare and military skills from Iraq to Syria.

ISI intelligence-gathering started in early April 2011, but the organisation's predecessors already had significant experiences and information on the Syrian political, military and jihadist scenes as early as 2004.[29] At a later stage, in 2013 and 2014, these capacities developed further. ISIS commanders collected intelligence on powerful families, leading insurgent figures, influential tribal/clan leaders, topographic data and even 'dirt' on rebels in targeted areas liberated from Assad's forces.[30] The organisation pursued an 'al-Saud style' of tribal infiltration by marrying into the influential families/clans to gain allegiances and intelligence.[31]

In terms of manpower, JN – still secretly affiliated with ISI – had about 30 fighters under its command in August 2011. During that month, Abu Muhammad al-Julani met with Zahran Alloush, the founder of the Army of Islam,[32] in al-Qalamun hills.[33] The man sent by the ISI leaders asked Alloush to join him in a new organisation. When asked about his manpower, al-Julani allegedly replied that he had only 30 fighters.[34] Alloush declined the offer, having already about 300 fighters under his command.[35] Bold, experienced and undeterred by their tiny size in Syria, ISI-JN still attempted to absorb already existing armed organisations. The JN leadership contacted Hassan Abboud (Abu Abdullah al-Hamawi) – the leader of *Ahrar al-Sham* (Freemen of the Levant) and Abdul Qadir Salih (Hajji Mari') – the commander of *Liwa' al-Tawhid* (Monotheism Brigade). These two were among the largest armed opposition organisations operating in north-central and northwest Syria. The absorption attempts by JN were all failures, however. They would later result in brutal hostilities. Despite that, by the summer of 2012, JN had multiplied its manpower by almost 100 times, mainly via recruiting individuals and factions as opposed to absorbing large organisations. Some of the estimates even exceeded that. According to a Free Syrian Army (FSA) commander, JN's manpower ranged between 6,000 and 10,000 men in November 2012.[36] In April 2013, ISI leader, Abu Bakr al-Baghdadi, decided to 'merge' the JN and the ISI to become ISIS. Al-Julani rejected the merger and, as a result, the JN lost most of its manpower and well-over half of its territories, arsenals, buildings, camps and other valuable resources in Syria to the newly born ISIS.[37] By mid-2014, the CIA estimated IS manpower to range between 20,000 and 31,000 fighters.[38] However, that estimate included both Iraq and Syria. In January 2020, the Syrian Observatory for Human Rights (SOHR) published a study that claimed that the number of bodies of IS fighters in Syria alone numbered 37,707.[39] The claimed figure is likely to be at the very high-end of the estimates.[40] Put together with the estimates of surviving members in al-Hawl and other detention camps,[41] and the IS operating units in Syria in 2020 (*Wilayat al-Sham* or The Levant Province, especially in al-Badiya/Homs Desert),[42] this would mean that IS had about 50,000 members, only in Syria. This far-exceeds all credible estimates of ISIS/IS manpower in the war-stricken country.

ISIS only lasted between April 2013 and June 2014, before it became IS. During that period, the ISIS operations in Syria were not as thoroughly reported in their annual metrics as those of ISI/ISIS in Iraq.[43] However, between 2013 and 2020, ISIS/IS in Syria reported conventional, guerrilla and/or terrorism operations in all of the 14 Syrian governorates. By the end of 2019, the organisation had controlled territories or denied them to other forces in 13 out of the 14 Syrian governorates (all except Tartus). This was done despite a very low fighter-to-space ratio and a relatively limited manpower and firepower. By 2014, ISIS/IS in Syria had a complex conventional, guerrilla and terrorism arsenal that later allowed for territorial control or territorial denial of about 50 per cent of Syria's territories in 2015.[44] This arsenal was the organisation's most sophisticated in terms of quality and quantity of weapons compared to anywhere else, except Iraq.

In terms of conventional capacities, IS constantly raided the regime's military bases and weapon depots, including the 17th Division military base, the 93rd Brigade military base and al-Tabqa military airbase (all in Raqqa), Ayyash weapons depot (Deir Ezzor) and other major depots (in Deir Ezzor and Homs, especially in Palmyra).[45] IS had looted more than 200 tanks in Syria by the end of 2015.[46] These included a *minimum* of 79 T-55 tanks, 25 T-62 Tanks and 19 T-72 tanks.[47] The organisation had more than 60 armoured and infantry fighting vehicles in its conventional arsenal in Syria.[48] Many of the captured tanks and infantry fighting vehicles were modified by ISIS/IS in its own local workshops. The two workshops in Raqqa Governorate (in Thawrah Industrial Facility near al-Tabqa military airbase) and in Deir Ezzor Governorate (Tracked Armoured Facility) were especially significant.[49] The two workshops had overhauled, modified and upgraded multiple specifications of (mainly) Soviet and Russian tanks, infantry fighting vehicles and other conventional armoured and artillery pieces. The altered specifications included upgrading or modifying the armour, as well as the principal and secondary armaments of BMP-1 infantry fighting vehicles, T-55, T-62 and even the relatively more advanced T-72 tanks.[50] IS workshops have also converted trucks into weapons platforms. One example was mounting a 122mm howitzer (D-30) on a flatbed Toyota truck. Another example was mounting BMP-1 turrets on Toyota Land Cruisers.

ISIS/IS had over 120 artillery pieces in Syria.[51] Those ranged from 60mm light mortars to 155mm M198 Howitzers.[52] The organisation exhibited over 50 unguided multiple rocket launchers in the media releases of its Syrian Provinces.[53] These included tens of Soviet-made BM-1 Grad rockets (122mm) and Chinese-made Type-63 rockets (107mm).[54] As elsewhere, ISIS/IS Provinces in Syria had limited, short-range mobile air-defence capabilities. The organisation has shown over 100 anti-aircraft guns of different types. These ranged from tens of KPV heavy machine guns (14.5 mm) to Type-65 (37mm) and AZP S-60 (57mm) anti-aircraft artillery.

The Raqqa Province of IS exhibited at least 4 ZSU-23 (Shilka) self-propelled, anti-aircraft gun with four autocannons in its photographic reports. The most prolific weapon in the anti-aircraft arsenal of ISIS/IS in Syria was the ZU-23 (23mm). It was exhibited over 50 times by ISIS/IS Syrian Provinces, including in the Provinces of Raqqa, Aleppo, al-Baraka (al-Hasaka) and al-Khayr (Dier Ezzor). Finally, IS had unconfirmed numbers of different types of MAN-PADS in Syria. The MANPADS arsenal included the old Soviet-made SA-7 (Sterla family), the more advanced, infrared-homing S-16 and SA-18 (Igla family),[55] Chinese-made FN-6 and North Korean-made HT-16.[56] During the attack on the 66th Brigade of the Syrian army in Hama Governorate, IS Hama Province also captured an SA-3 GAO surface-to-air missile system.[57]

In terms of airforce, IS managed to capture more than 15 MiG-21 jet fighters and interception planes exhibited mainly by the Raqqa Province media office. The organisation also looted an unknown number of AA-2 (Atoll) air-to-air missiles. These were used later in ground-fighting as variants of IEDs.[58] According to the commander of Raqqa's Military Council in the Syrian opposition, Airforce Colonel Mu'taz Raslan, IS was able to conduct at least two short-range reconnaissance air-sorties at low altitude in Raqqa Governorate using two of the MiGs.[59] Despite recruiting Syrian and non-Syrian pilots and maintenance crews, the organisation had no capacity to conduct airstrikes. Still, it tried. IS Raqqa Province fighters forced some of the prisoners who worked as a ground crew in al-Tabqa airbase to conduct maintenance tasks in September 2014.[60] However, all of the captured MiGs were abandoned or destroyed at a later stage.

In addition to the conventional capacities, by late 2014, IS provinces in Syria owned a sophisticated guerrilla arsenal. As a reflection of its capacities, the Conflict Armament Research (CAR) group has documented 562 weapons and 11,816 units of ammunition that belonged to IS in Syria between July 2014 and November 2017.[61] This sample featured 18 types of weapons, all of which are commonly used in guerrilla warfare. The sample included handguns, assault rifles, light, medium and heavy machine-guns, sniper rifles, grenade-launchers and unguided rocket-launchers.[62] However, the sample is by no means close to an exhaustive list. IS had looted guerrilla/infantry weapons from the 17th Special Forces Division in Raqqa in July 2014 that included over 30 light and medium mortars, over

100 heavy machine guns (12.7mm and 14.5mm calibre), hundreds of RPG-7 grenades, over 3,000 variants of AK assault rifles, over 1000 handguns and hundreds of silencers.[63] Finally, Raqqa, al-Baraka and Homs (al-Badiya) IS Provinces in Syria exhibited various type of ATGMs, including Soviet-made, wire-guided Konkurs and the more advanced and expensive, laser-guided Kornets. This is in addition to a large number of Metis-M ATGMs.

In terms of terrorism capacities, ISI had about five years' worth of experience in IED-manufacturing and IED-intensive warfare when it sent the founding squad of the JN in August 2011.[64] ISI transferred its experiences in small-scale production of suicide belts and vests for individuals, and of fitting passenger cars and motorcycles with explosives to produce S/VBIEDs and S/MBIEDS in small work-shops to the JN. For its part, the JN developed its own capacities to transform urban terror tactics into effective battlefield ones. The JN and then ISIS/IS relied on SVBIEDs as battlefield weapons, but also innovated variants of VBIEDs ranging from radio-controlled toy cars to armoured vehicles and self-propelled artillery to deliver the explosives to the target.[65] As in IS Provinces elsewhere, these variations of VBIEDs in Syria functioned more like guided rocket bar-rages and artillery in conventional military tactics. This is in addition to their more common usage in terrorism tactics. IS has also looted over 1,000 handguns and over 300 silencers from the 17th Division in Raqqa in July 2014 after capturing it from the regime's forces in a blitzkrieg-like operation. Many of these 'silenced handguns' were used in assassination operations conducted in both rebel- and regime-held territories.[66]

As in the other countries, IS in Syria developed the 'knowl-edge transfer' pillar via various commanders and platforms, including training camps. More than 20 makeshift training camps were identified in Syria by late 2014.[67] By mid-2015, IS had at least nine main camps in Syria, where a mix of conventional, guerrilla and terrorism tactics were taught. Almost all of these 9 camps were located in northern and northeastern Syria; in al-Raqqa Governorate (near Suluk), in Deir Ezzor countryside and in Aleppo Governorate (near Tadif).[68]

Overall, in about two years, ISI's main affiliates[69] and successors in Syria (JN and ISIS) had transformed from a small, platoon-sized band of 30 fighters in mid-2011 to two military organisations with conventional, guerrilla and terrorism capacities by mid-2013. After the establishment of IS in June 2014, the organisation expanded its territorial control/area-denial capacities, developed its conventional armour offensives, siege and urban warfare tactics and upgraded its sniper operations as well as its artillery and rocket-fire support without line-of-sight availability. IS has also managed to organise large-scale logistics movements between its Syrian and Iraqi 'Provinces'. IS in Syria used unmanned aerial vehicles (or drones) for the purpose of forward observation, surveillance, reconnaissance, propaganda, combat and other roles. All of these hybrid capacities and developments were utilised by IS to fight almost all state and nonstate armed actors operating in Syria and to terrorise large segments of the local population, both Sunni and non-Sunni. The organisation succeeded in occupying, denying or exercising influence in about 50 per cent of Syria's territories, before the overwhelming majority of these areas were liberated from IS by the combined (and often conflicting) military efforts of state and nonstate armed actors in 2019.

The capital and the governorate: the battlefronts of Raqqa (April 2013–October 2017)

On 4 March 2013, a coalition of Syrian armed opposition organisations managed to oust the regime's forces from Raqqa City. The city became the first provincial capital to be liberated from Assad's forces. The rebels' coalition included units from several Free Syrian Army factions, *Ahrar al-Sham* and, also, the JN. The gained basic freedoms did not last long in the city, however. About a month after the liberation, ISIS was established, in an attempt to merge ISI in Iraq with the JN in Syria. The JN leadership rejected the merger. From the end of April 2013, the majority of JN fighters based in Raqqa switched their allegiance to ISIS, including the JN *emir* (commander) of the city, Abu Saad al-Hadrami (Ibrahim Sa'id al-Abdullah).[70] In about a week, ISIS absorbed factions from the JN and others, and also recruited individual fighters and factions from other organisations[71] via a mix of earlier infiltration, intense propaganda, selective incentives and better intelligence than the Syrian rebels. The capacities of ISIS in Raqqa were thereby upgraded from almost nothing to one of the most military capable organisations in the city.

On the offensive: the expansion of IS in Raqqa

To occupy the city and the whole Governorate of Raqqa, ISIS had to fight all of the Raqqa-based rebel and regime forces in three campaigns. The first two campaigns mainly targeted the rebels; between May and September 2013 and in January 2014. The third campaign mainly targeted the regime's forces in the summer of 2014. By September 2014, IS had occupied the whole governorate through a mix of terrorism, guerrilla and conventional tactics, despite being initially outnumbered and outgunned by both decentralised rebels and the relatively centralised regime's forces.

During the first campaign/series of battles, ISIS' *modus operandi* in Raqqa was similar to the one employed later in some of the Iraqi towns and by IS Cyrenaica Province in Derna (between January and June 2015).[72] The *modus operandi* was successful in the case of Raqqa compared to Derna, however. ISIS/IS initiated a sustained terror campaign in Raqqa that carefully selected its targets between May and August 2013. The organisation kidnapped and assassinated tens of civil society leaders and opinion-makers as well as rebel commanders.[73] On 13 August 2013, IS managed to destroy the headquarters of *Ahfad al-Rasul* (Descendants of the Messenger) – one of the FSA-aligned, Raqqa-based brigades – after breaching its security provisions with a covert VBIED.[74] That was followed by three other VBIEDs detonated in other *Ahfad* and FSA positions in the city. A day later, *Ahfad al-Rasul*, one of the main rebel organisations in the governorate, withdrew from Raqqa City. ISIS continued 'creeping' in on other neighbourhoods of Raqqa City and 'softening' other rebel armed units. The organisation detained its former *emir* of Raqqa City, Abu Saad al-Hadrami, in September 2013.[75] By doing so, it pre-emptively decapitated and botched an attempt to form an anti-ISIS coalition in Raqqa, composed of armed units from five different organisations and factions.[76] By September 2013, ISIS had become the most dominant armed organisation in Raqqa by relying almost exclusively on urban terrorism and intelligence-like infiltration and pre-emption operations.

Across the liberated parts of northern and central Syria, from Latakia to Deir Ezzor, ISIS was using similar terror tactics to those used in Raqqa. The organisation committed a sustained series of assassinations, kidnappings and bombings in 2013 against rebel forces in an attempt to 'soften' and terrorise them into submission, and 'creep in' on rebel territories. These tactics, however, had a similar Ramadi- and Derna-like effects.[77] They mobilised almost all of the northern, and some of the southern, rebel groups against ISIS by January 2014. The coordinated offensive of January 2014 involved units from several FSA factions (including the almost 10,000-strong *Liwa' al-Tawhid*), as well as units from *Ahrar al-Sham*, *Jaysh al-Islam* (Army of Islam) and the JN. These relatively well-coordinated attacks forced ISIS units to retreat from most of its northwestern positions in parts of the Latakia, Idlib, Aleppo and Hama governorates. As a result, three main ISIS columns were retreating via eastern rural Aleppo governorate towards the Raqqa governorate. The first ISIS column retreated via al-Atarib in central-west Aleppo and included ISIS units that fought in Latakia, Idlib and Aleppo. The second ISIS column retreated via Khanasir in central-south Aleppo. The third ISIS column retreated via the far-south of Aleppo. Abu Bakr al-Baghdadi (former ISI/ISIS/IS leader), Abu Ayman al-Iraqi (former commander of Latakia units) and Abu Muhammad al-Adnani (former IS spokesperson) were all retreating in that third column.

In Raqqa, the evening of 2 January 2014 witnessed a meet- ing between the local leaders of the *Ahrar* (Abu Haydara), the JN (Abu al-Abbas) and the FSA's Revolutionaries of Raqqa Brigade (Abu Issa). The commanders hatched a battleplan to remove or destroy ISIS in the city.[78] When the second-battle phase unfolded in Raqqa on 4 January 2014, ISIS' manpower in the whole gover- norate ranged between 1,500 and 2,000 fighters (aiming to control roughly 7,500 square miles).[79] In the City of Raqqa, ISIS had about 500 fighters.[80] The combined forces of the *Ahrar*, the JN and the remaining units of the FSA factions in the city were estimated to exceed the 2,500 fighters.[81] Hence, ISIS was outnumbered five-to- one before its reinforcements arrive. The overwhelming majority of ISIS manpower, about four platoon-sized guerrilla formations (GFs), was concentrated in the city centre in and around *Tall Abyad* (White Hill) Street. *Ahrar al-Sham*'s units and the remaining 'Revolutionaries of Raqqa Brigade' fighters were concentrated in the western neighbourhoods of Raqqa, whereas the JN units were concentrated in the east of the city.[82] A coordinated JN, *Ahrar* and FSA attack started on 4 January 2014 on all six ISIS main posi- tions in the centre of Raqqa.[83] Outnumbered and surprised by the unexpectedly coordinated attacks, all ISIS formations retreated to the Raqqa Governorate HQ and the streets around it by 6 January 2014. A few squad-sized units also retreated to the north of the city. At this point, ISIS looked like it was losing the 2014 battle of Raqqa.

Two developments changed ISIS' fate in the Raqqa battle of 2014. First, an ISIS reinforcement of about 30 soft-skinned technicals and 4x4 vehicles arrived in the early hours of 8 January 2014.[84] The irregular column was composed of retreating ISIS fighters from the governorates of Idlib, Aleppo and Deir Ezzor. Somehow – either through deception or infiltration – this convoy passed through the JN checkpoints in the east of the city without a fight.[85] The arrival of reinforcements strengthened an ongoing counteroffensive on the *Ahrar's* positions in the west of the city.[86] By 9 January 2014, ISIS was victorious in the centre and the west of Raqqa City. Both the *Ahrar* and the FSA units were retreating northwards towards the town of Tell Abyad near the Turkish borders. Nearby, ISIS had already absorbed some of the FSA factions operating in northern Raqqa Governorate such as al-Hamza Brigade and *Sawa'iq al-Rahman* Battalion.[87] Both factions were in strategic positions in rural Raqqa and ambushed the *Ahrar* units fleeing Raqqa City, with the help of some of the retreating ISIS units form eastern rural Aleppo. The ambush happened near al-Kantari area about 30 miles north of Raqqa City. One ISIS Tunisian commander was heard over a wireless device ordering his fighters to 'take no prisoners from the *sahwat* [awakening councils]'.[88] After that, ISIS seized the momentum of their counteroffensive to storm and occupy the town of Tell Abyad and capture more territory in Raqqa Governorate.

Back in the eastern neighbourhoods of Raqqa City, the tenacity of ISIS' counteroffensive caught the rebels off guard, despite earlier warnings. This was the second development that helped turn ISIS fortunes in Raqqa City. The organisation ultimately focused their counteroffensive on the JN forces during the late night of 8 January 2014. And in a show of force, ISIS announced the decapitation of its detained former *emir* of Raqqa, Abu Saad al-Hadrami. The latter defected back to the JN (after previously defecting to ISIS) and attempted to organise an anti-ISIS resistance campaign in Raqqa. After four days of fighting in the eastern neighbourhoods and the suburbs of the city, the JN units retreated southwards and then south-westwards towards the town of al-Tabqa. Unlike the *Ahrar*, the JN destroyed all of its heavy weapons including two T-55 tanks and several heavy mortars (120mm), so that they do not fall into the hands of ISIS fighters.[89]

By 13 January 2014, ISIS was in relative control of Raqqa City and continuing its offensive to occupy the rest of the governorate.[90] As in the north of Raqqa Governorate, the organisation seized the momentum of its counteroffensive in the southwest. It followed the JN to al-Tabqa, stormed its main checkpoint with an SVBIED and apparently infiltrated the JN units in al-Tabqa with a security/spy detachment (*mafraza amniya*).[91] The JN units started retreating from al-Tabqa. They were rapidly followed by an FSA unit from Owais al-Qarni Brigade.[92] In the southeast of Raqqa governorate, the town of Karamah was occupied without fighting.[93] The town of Ma'dan in the far southeast of rural Raqqa was also occupied by ISIS after some resistance from *Ahrar al-Sham* fighters there. However, they surrendered to ISIS forces after the organisation besieged and shelled the town for a few hours. By 15 January 2014, ISIS was decisively victorious against all rebel forces in Raqqa Governorate.

Taking on the regime

The third campaign/series of battles for Raqqa Governorate was against the regime's forces. After consolidating relative control in most of the governorate, capturing rebels' arms, ammunitions and resources, as well transferring some of the looted weapons from its campaigns in Iraq; ISIS targeted the regime's powerbases in Raqqa Governorate. The hard-targets were the headquarters and the military base of the 17th Special Forces Division north of Raqqa City;[94] the headquarters of the 93rd Armoured Brigade in 'Ayn 'Issa (Eye of Jesus); and al-Tabqa Military Airbase in al-Tabqa.

On the late night of 23 July 2014, ISIS (now IS as of 29 June 2014) started with the 17th Division, a two square-mile military base encompassing the headquarters of twelve battalions, located north of Raqqa City.[95] The IS attacking force of about 300 fighters was divided into six platoon- and small company-sized guerrilla (GFs), suicidal guerrilla (SGFs), armoured and artillery formations.[96] With about 800 Special Forces' officers and soldiers left in the division, IS was on the offensive while being slightly outnumbered about 2.5-to-1. Despite the full siege laid on the division using at least 15 tanks,[97] IS units initially attempted to storm the base from only three sides. The south and the southeast of the division were attacked by two SVBIEDs, followed by two failed infiltration attempts by SGFs. The failure was due to strong resistance and anti-aircraft autocannons (23mm) and heavy machine guns fired from elevated grounds.[98] The western edges of the division were shelled with heavy artillery, likely a 130mm towed Type-59 field gun and medium mortars (82mm). From the northeast, where there was no attack, about 30-to-40 IS fighters managed to infiltrate the division in the early hours of 24 July 2014.[99] This sparked the beginning of the end of the battle, despite the relatively limited number of infiltrators and about thirteen air-raids by the regime's air-force on IS positions in Raqqa.[100] The infiltrating, platoon-sized formation managed to wreak havoc inside the Division with two SVBIEDs detonated by two Saudi suicide-bombers (Abu Suhayb and Khattab) in the early hours of 24 July 2014.[101] IS units also managed to take over the 'chemical battalion' building and the sugar factory – located on elevated ground, and used these locations for sniping and light-artillery shelling.[102] The fighting continued for the rest of the day and by the night of 24 July 2014, IS was clearly victorious. The 17th Special Forces Division that had withstood the attacks of armed opposition factions for more than a year was taken by IS in about two days of intense fighting.[103]

Seizing the momentum of its victory over the 17th Division and the nearby 121st Artillery Regiment in Milbiyah (in al-Hasaka Governorate),[104] the high morale of its fighters and the quantity of captured weapons from both the division and the regiment; IS developed its offensive to control the rest of Raqqa Governorate. The organisation attacked the 93rd Brigade in 'Ayn 'Issa (Eye of Jesus) – about 30 miles north of Raqqa City – on the night of 6 August 2014. Hundreds of soldiers fleeing the 17th Division either reached the 93rd Brigade or were still heading there. Some of them were ambushed on their way north by IS units.[105] Despite the availability of enough heavy weapons for conventional land warfare – including a number of T-55 and T-62 tanks, 130mm artillery pieces and other pieces of lower calibre – IS initiated the attack on the Brigade with three SVBIEDs from three sides.[106] Like the attack on the 17th Division, this was followed by rapid advancement of guerrilla formations and an infiltrating unit.[107] The organisation did not modify the offensive battleplan of the 17th Division, despite the availability of more man- and firepower. It still succeeded. When IS units looked clearly victorious over the 93rd Brigade, an estimated 36 soldiers and 15 IS fighters were reportedly killed in combat.[108] The organisation captured over a dozen T-55 tanks and more than ten 130mm artillery pieces from the brigade.[109] The next and final target to execute the strategic plan of occupying Raqqa Governorate was al-Tabqa military airbase. The latter was the sole airbase of the regime's forces in Raqqa and their last stronghold in the governorate. The airbase is about eight square-miles near the town of al-Tabqa, with facilities including hardened aircraft shelters and ordnance storages for, at maximum, 55 planes.[110] On 10 August 2014, about 500 IS fighters started harassing the airbase in which well-over 1,000 soldiers and officers were loosely besieged.[111] The heavy fighting took place between 19 and 24 August 2014. On 19 August 2014, IS escalated with two SVBIED-led attacks, followed by several platoon-sized GFs and SGFs advances.[112] At least two suicide vests were detonated during that attack, but the infiltration attempt failed.[113] The escalation re-alerted both the Assad regime and international powers to the gravity of the situation. Over the next four days, the regime's forces counterattacked with at least fifteen air-raids on IS positions.[114] They were allegedly assisted by US intelligence on IS positions. 'The Americans passed coordinates to both Iraqi and Russian security institutions', said a former Syrian officer who was involved in combating IS.[115] After three other failed infiltration attempts, IS units managed to break through after a fifth SVBIED-led attack. The units infiltrated the airbase for the first time on the late night of 23 August 2014, mainly through sustained usage of SVBIEDs and SGFs.[116] At least fourteen suicide vests and SVBIEDs were detonated throughout the four-day battle.[117] By the end of 24 August 2014, IS was victorious in the battle for al-Tabqa airbase, as well as the battle for Raqqa Governorate.[118]

Defences and counteroffensives: endurance of IS in Raqqa

Before the coalition-led operation to liberate Raqqa in November 2016, a number of attempts were made to recapture the governorate. These attempts included the June 2016 failed campaign of the regime's forces known as 'To Raqqa'. The campaign was well-timed to coincide with another campaign aiming to liberate the town of Manbij and the areas around it from IS by the coalition forces and the SDF (May to August 2016). IS, however, was unexpectedly able to fight on two different battlefronts in northern Aleppo and central-southern Raqqa with limited manpower. Despite the good timing and other helpful factors, the campaign ended disastrously for the regime and as a victory for IS Raqqa Province. The campaign had showed up several weaknesses in Assad's forces. The regular army suffered from a combination of weak intelligence, surveillance and reconnaissance, limited combined arms coordination, inability to counterattack or improvise and rapid fragmentation of its units under IS tenacious counterattacks. IS was relatively better in all of the above-mentioned dimensions, despite being weaker in others such as manpower, firepower, airpower, air-defence, popular-support and (lack of) state-sponsorship. The three weeks of fighting in June 2016, however, wore down IS Raqqa Province and exposed some of its tactical innovations and defensive operational plans.

A relatively successful coalition-led operation to liberate Raqqa Governorate started on 6 November 2016 in the town of 'Ayn 'Issa, near the 93rd Brigade military base. It did not reach the City of Raqqa, only 30-miles away, until seven months later on 6 June 2017. The battle to take the City of Raqqa from IS was the final phase of a larger campaign, codenamed 'Operation Wrath of the Euphrates'. The strategic objective of that campaign was to isolate and liberate the city, after capturing the whole governorate from IS forces. The airstrikes, the strategic support and the planning of the campaign were led and conducted by the United States and its western allies. The ground troops of the campaign were primarily composed of a coalition of 18 armed nonstate actors led by the People's Protection Units (YPG),[119] and represented by the broader multi-ethnic coalition of Syria's Democratic Forces (SDF).[120]

The initial coalition battleplan was composed of two phases: to seize the towns and villages around Raqqa City and then advance on the city from three fronts (north/northeast, southeast and west). A mix of IS innovative tactics and strategic shifts applied in restless manoeuvres and relentless counteroffensives led to delays and changes to the battleplan, which became a five-phase plan stretched over four months and two weeks (6 June to 20 October 2017). Due to a bloody stalemate in the city centre, the urban battle-outcome included a surprising compromise reached by mid-October 2017. It allowed hundreds (and perhaps thousands) of IS fighters and their family members and loyalists to leave the city with their weapons and ammunitions.[121]

By 24 June 2017, Raqqa City was completely encircled by the SDF and the Coalition forces. IS was on the defensive while being outnumbered 10-to-1, at least.[122] The SDF and their allies managed to mobilise over 35,000 soldiers and fighters from 18 armed organisations operating in Raqqa and nearby governorates.[123] Many of the fighters belonged to the Kurdish-dominated YPG and YPJ[124] units. Arab units were also present. They were mainly composed of tribal fighters such the al-Sanadid (The Brave) Forces (clans from the Shammar tribe) and the Elite Forces (clans form the Shu'aitat tribe). Raqqa's liberating forces also included US and western army units, estimated to range between 3,000 and 5,000 soldiers.[125] They included 500 US Special Forces who were involved in the ground combat.[126] This is in addition to British, French and German special forces units, military advisors and close air-support, all operating as partners in the CJTF-OIR. During the campaign, the US artillery battalion fired over 40,000 shells (including 34,033 155mm heavy artillery rounds). To give a comparative reference-point, this was more than all the shells fired during the entire 2003 invasion of Iraq.[127]

Throughout June 2017 – the first month of the battle – the US airforce conducted heavy airstrikes on IS positions in Raqqa City. Official CENTCOM data shows a record of 644 air and artillery strikes in Raqqa Governorate aiming at 1,475 targets.[128] Overall, over 4,400 munitions were fired to degrade and destroy IS positions in Raqqa in the 30 days of June 2017. The organisation's limited air-defence systems could not respond. Under the airstrikes, the SDF attacked from the north (targeting the 17th Division HQ), northwest (al-Andalus neighbourhood), southeast (al-Mashlab neighbourhood) and west (al-Sibahiyyah suburb). In June 2017, IS managed to defend most parts of these neighbourhoods, mainly through its version of 'combined arms:' IEDs, S/VBIEDs, sniping- and tunnel-intensive tactics and drone attacks as detailed below. When the SDF managed to take parts of the 17th Division and al-Mashlab by 8 June 2017, IS' GFs and SGFs aggressively counterattacked and recaptured them. By 10 June 2017, the SDF managed to advance in the western suburbs of al-Rumaniyyah and al-Sibahiyyah, despite at least two SVBIEDs detonated by two Emirati foreign fighters (Abu Firas and Abu 'Awf) and a failed counteroffensive by three platoon-sized GFs.[129] By mid-June 2017, intense urban battles were raging in all of Raqqa's neighbourhoods, including the heavily fortified city centre where the elite of IS' GFs and sniper formations were concentrated.[130] By the end of June, IS units had launched counteroffensives in all lost neighbourhoods. The outcomes of these counteroffensives were mixed. They failed in al-Mashlab and Batani in the east, and in al-Qadisiyyah, al-Rumaniyyah and al-Sibahiyyah in the west. However, the organisation successfully defended the city centre and almost all of the northern neighbourhoods. IS finished the first month of the battle by recapturing parts of the Industrial (*al-Sina'iyyah*) neighbourhood in the southeast of city, leading the counteroffensive with a wave of SVBIEDs.

The tactics used by the warring sides during July and August did not differ much from June. However, IS relied less on drone-strikes and more on a network of tunnels that it had been digging and developing as early as January 2015.[131] The organisation used the tunnel networks against the SDF in Batani (earlier in June 2017), the Old City, Hisham Ibn Abdul al-Malik and Nazlat Shihada neighbourhoods.[132] The SDF units would believe they had secured a neighbourhood, before an IS' GF or an SGF emerges from behind their lines using a tunnel. This occurred over ten times between June and August 2017.[133] Overall, the battle outcomes in July and August were not too far from the June outcomes. Despite the slow advance of the SDF under heavy US precision strikes in the south, east and west of Raqqa City; several neighbourhoods kept exchanging hands between the SDF and IS. This included parts of the Old City (centre), al-Barid (east of centre) and Hisham Ibn Abdul al-Malik (southeast of the centre). By the end of August, however, IS had lost almost all of the southern neighbourhoods and suburbs of Raqqa.

In September, the SDF was slowly but steadily advancing on the Old City and, generally, near the south of the city centre. This included the heavily IED-fortified Thakana neighbourhood. The advance of the SDF in these areas was assisted by close-air support, involving a number of US AH-64 Apaches attack helicopters. By early September, it became clear that IS had lost its command and control structures.[134] Raqqa Province fighters were operating as autonomous units in different neighbourhoods of the city. They still managed to launch 'combined armed' tactics. Also, IS SGF units continued using hidden tunnels that led them behind the

SDF lines, while being disguised as civilians or as SDF militiamen. Hence, the SDF was too quick to declare victory in Raqqa on 22 September 2017.[135] Four days later, on 26 September 2017, an IS platoon-sized SGF disguised as an YPG unit managed to infiltrate the SDF positions near Rumayla (northeast). It succeeded in killing at least 28 SDF fighters there.[136] In the same week, IS units managed to launch counteroffensives in three south-eastern neighbourhoods (al-Mashlab, al-Sina'iyya, and Batani). These neighbourhoods had all been previously liberated by the SDF.[137]

The Raqqa battle continued into October, mainly in city centre. Unexpectedly, it reached a compromise that raised eyebrows. A fierce battle was raging in a city centre triangle of the national hospital, the stadium and al-Na'im (bliss or heaven) roundabout. The latter was dubbed the 'hell (jahim) roundabout' due to the intensity of the fighting there.[138] IS units managed to hold on to the three locations, in addition to the grain silos area (north of Raqqa city centre and just south of the 17th Division HQ). Earlier in July 2017, the CJTF–OIR chief commander, Lieutenant-General Stephen Townsend, stated that IS units in Raqqa had only two options: 'to surrender or be killed'.[139] IS Raqqa Province gave one option to the SDF and the Coalition. It was outlined in a statement published in the 84th issue of al-Naba' newsletter dated 8 June 2017. This option was clearly indicated in the title of the statement: 'We Either Eradicate the Polytheists or Die Trying: There is no Third Choice'.[140] Instead, both IS and the SDF had to negotiate. This was partly due to war-fatigue and an attempt to avoid further losses among their fighters. It was also due to how IS fought — the mix of innovative and 'combined arms' tactics, strategic shifts, aggressive and tenacious counteroffensives, and relative combat effectiveness in a context where a strategic victory was unattainable.[141]

On 15 October 2017, IS Raqqa Province fighters along with their weapons and ammunition, their family members and loyalists were allowed to withdraw from al-Na'im roundabout and the area around it in a four-mile convoy of about 50 trucks, 13 buses and more than 100 IS vehicles.[142] The actual number of IS fighters in the convoy remain debatable, and the difference between the estimates is significant. According to the SDF, less than 300 IS fighters were on the convoy.[143] A YPG defector stated (a less-likely estimate) of more than 3,000 IS fighters.[144] Five days later, on 20 October 2017, the SDF officially declared victory in Raqqa City.[145] To celebrate, YPG and YPJ fighters raised a huge picture of Abdullah Öcalan, the founder of the Kurdistan's Workers Party (PKK), in *al-Na'im* roundabout in the city centre of Raqqa.[146]

How ISIS/IS fights in Syria: battlefronts analysis

As outlined above, ISIS/IS pulled off several military upsets in Syria, particularly in Raqqa. Like the cases of Iraq, the use of the innovative tactics and overall combat effectiveness of the organisation resulted in surprising operational victories in the short-to-mid-term. The following section analyses some of the tactical and operational patterns employed by ISIS/IS in Syria. It is based on quantitative data and qualitative surveying of the details of IS operations in Raqqa, Deir Ezzor and other Syrian governorates between late 2013 and early 2020.

The Raqqa battlefronts

The offensive and defensive battles in Raqqa Governorate provided unique insights into how IS developed its ways of warfare in urban and rural terrains. Prior to April 2013, before the establishment of ISIS, the ISI/JN fought in a coalition to capture Raqqa City. None of the Syrian rebel organisations were aware of ISI's future battle-plans (or aware of its existence among them given the JN's denials). After April 2013, ISIS used a mix of urban terrorism and guerrilla warfare tactics to 'soften' other armed organisations and then 'creep' on their territories. It liquidated other armed organisations and leading activists and consolidated its power in Raqqa City. This was followed by the attacks on the Assad regime's powerbases in Raqqa Governorate. Similar to Iraq, ISI/ISIS' used a three-phase, operational-level *modus operandi* in Syria (soften-creep, coalition-build and liquidate-consolidate or SCCLC).[147] It aimed to occupy Raqqa City via this *modus operandi* between April and September 2013. It was almost the same *modus operandi* successfully employed in Fallujah (January 2014) and Mosul (June 2014); and unsuccessfully employed in Ramadi (October 2014) and Derna (by an ally of IS before November 2014 and afterwards by IS).

In the battles to remain and endure in Raqqa Governorate (6 November 2016 to 20 October 2017), IS has mainly relied[148] and reported on fourteen categories of tactics illustrated below in Table 3.1 (overleaf).[149]

Table 3.1 IS Reported Categories of Tactics in Raqqa Governorate, November 2016 to October 2017

Category of Tactics	Reported Usage
Static IEDs	362
SVBIEDs' Variants	224
Unguided Rockets	131
Suicidal Guerrilla Formations (SGFs)	105
Sniping	90
Guerrilla Formations (GFs)	72
Artillery	54
ATGMs	41
MANPADS	21
Anti-aircraft Autocannons	17
Tunnels	17
VBIEDs' Variants	14
Close-quarter Assassinations	11
Total	**1159**
Drones	**+200**

Shelling by blowing: SVBIEDs and other IED-intensive tactics

The reliance on (variants of) SVBIEDs – an urban terrorism weapon – as a conventional battlefield weapon remained the hallmark of IS tactics in Syria in general, and during the battles of Raqqa in particular. The organisation has used SVBIEDs and SVBIED-intensive tactics effectively in offensive, counteroffensive and defensive operations throughout Syria. Overall, IS in Syria has used at least eight types of SVBIEDs, reported in Raqqa, Aleppo, Homs and Deir Ezzor Governorates:[150]

1) Soft-skinned, covert SVBIEDs emulated civilian vehicles in order to creep on designated target(s).
2) Armoured, covert SVBIEDs also looked like civilian vehicles to confuse the SDF and the Coalition's aerial reconnaissance, while approaching their targets. These vehicles were internally armoured.

3) Armoured, camouflaged SVBIEDs would have the same colour as their surrounding environments (whether rural, urban or desert). They were also designed to confuse the SDF and the Coalition forces in the different theatres of Raqqa Governorate.

4) Externally up-armoured SVBIEDs are a fourth type used in Raqqa and Deir Ezzor Governorates.[151] IS primarily used them in the outer defensive lines in the governorates around Raqqa City suburbs. Like in the battlefronts of Mosul and Sirte, these specific SVBIEDs were effective in slowing down the advancing forces.

Within the armoured category, IS produced and utilised two innovative designs of SVBIEDs during the battles of the Raqqa and Deir Ezzor governorates:

5) Rocket-launcher-fitted SVBIEDs;
6) Machine-gun-fitted SVBIEDs.[152]

Both of these designs are manned by two suicide fighters, a driver and a gunner. The first suicide-fighter drives and detonates the vehicle when it eventually reaches the target. The second suicide fighter suppresses potential enemy fire to increase the possibilities of reaching the target.

The seventh and eighth types of SVBIEDs used in Raqqa Governorate[153] were:

7) Loaders-SVBIEDs; and
8) Bulldozers-SVBIEDs.

These were used against targets surrounded by large defensive structures, such as steel barriers and earth berms. These two designs were almost copycats from designs used in combat during the battle of Ramadi in May 2015.[154]

Two final observations should be mentioned about IS' SVBIEDs in Syria. First, the organisation used the abovementioned types in both urban and rural environments, almost equally. About 56 per cent or 126 of the SVBIEDs reported and recorded in the dataset were used in Raqqa City (June to October 2017).[155] The rest were reported in Raqqa Governorate, outside of the provincial capital. The organisation adapts and modifies its SVBIEDs tactics, regardless of the environment, theatre or target.[156]

The second observation is about the predilection towards, and the perceived effectiveness of, SVBIEDs within IS combat units. The organisation used SVBIEDs and SVBIED-intensive tactics in battles, when it did not *need* them. In the battle of the 17th Division and the 93rd brigade, IS was not only outnumbered (and therefore in *need* of manpower), but also had enough heavy weapons including a number of T-55 and T-62 tanks and heavy artillery pieces. Still, IS commanders preferred to use SVBIED tactics, or what Syrian jihadists refer to as *al-qasif bil nasf* (shelling by blowing-up).[157] This may be construed as a 'cult-like' degree of irrational reliance on SVBIEDs, even when alternatives are available. But it more likely shows that IS relies on tactics at which it *excels* (due to experience) as opposed to those which it is merely *capable of* (due to resources).

In terms of static IEDs, this category of tactics still topped the dataset in terms of reported quantities; a pattern observed in other battlefronts in Iraq, Libya, Egypt and elsewhere. The abovementioned figure in Table 3.1 lists only the reported usage. Hence, the figure is a conservative sample. IS had littered Raqqa City with static, covert IEDs. The qualitative range of tactics (and techniques) used and reported was extremely wide. It included rigging pressure plates, hiding under-floor motion-activated IEDs, concealing IEDs in ovens and drawers and attaching IEDs to pens, desk-chairs and corpses.[158] 'The terrorists [of IS] were trained to rig everything with explosives from pens to bulldozers. We suffered because of that. They don't teach us how to deal with this in al-Assad Military Academy',[159] said an FSA commander who fought IS in both Raqqa and Aleppo after defecting from the regime's army.[160] In the first two months of the battle of Raqqa (6 June to 3 August 2017), the CJTF-OIR spokesperson, Colonel Ryan Dillon, stated that more than 80 per cent of IS attacks against the SDF in Raqqa City were accomplished using IEDs.[161] In October 2017, Dillon stated that the usage of IEDs in the battle of Raqqa is even denser than other battles including those of Mosul, Ramadi and Fallujah.[162] Finally, IS also used parked, covert VBIEDs in Raqqa Governorate, leaving behind at least ten VBIEDs in SDF-captured areas, before detonating or attempting to detonate them.

Uncommon infantrymen: guerrillas, suicide guerrillas and snipers

In the battlefronts of Raqqa and in Syria generally, IS relied significantly on guerrilla formations (GFs) and suicidal guerrilla formations (SGFs or *inghimassiyun*). The GFs have the role of light infantry, sometimes motorised infantry and, rarely, mechanised infantry. The SGFs have similar roles comparable to conventional special forces or shock troops, except that they are also suicide bombers (usually wearing a suicide belt or vest). Throughout Syria, and especially in Raqqa and Deir Ezzor, IS GFs and SGFs sometimes exhibited several features of adequate tactical combat effectiveness. When they exhibited these features, the list included effective fire control, fire discipline (specifically when using a volley fire of unguided RPGs), fire and manoeuvre, interlocking fields of fire, and defence in depth (even creating 'hedgehog' defensive points[163] in the city centre of Raqqa).

The GFs and the SGFs of IS have done better in unconventional combined arms in Raqqa City. This was not just common guerrilla coordination such as coordinating between machine gun fire to suppress targets to enable other guerrilla units to attack with rockets and mortars. The tactics developed further to include the coordination between SVBIEDs, UCAVs/drones and GFs/SGFs, sometimes combined with aggressive swarming.[164] The GF/SGF units would usually capitalise on the havoc inflicted by the SVBIED(s), which were earlier guided to their target(s) by a UCAV/drone(s). Such unconventional combined armed tactics would sometimes be combined with swarming: engaging the SDF or other IS' enemies from all sides simultaneously. Frequently, one or two sides would be shelled by heavy artillery, whilst another side would be attacked by snipers and GFs. The attacked units would commonly withdraw or tactically regroup in the remaining one or two sides. IS units would then attack that side(s) with a barrage of SVBIEDs, followed by GFs or SGFs. These tactics were witnessed and repeated during IS operations in Raqqa, Deir Ezzor, Hasaka, Homs and Aleppo governorates.[165]

The GFs/SGFs in Raqqa (and in Deir Ezzor) had also shown adequate levels of manoeuvrability (relative to most of the regime's and rebels' forces) and some remote-firing innovations. This was especially the case in defensive and counteroffensive operations. Generally, IS fights small: squad- or platoon-size. When it conducted rare operations in company- or battalion-sized formations, it was usually during an offensive. The small size of the guerrilla formations facilitated rapid manoeuvrability, including shifting positions when flanked and relentless counteroffensives (causing damage regardless of the probabilities of success). When on the defensive, the GFs mainly fought as light infantry units in Raqqa and in Deir Ezzor, as the IS vehicles were relatively vulnerable to the OIR airstrikes.[166]

To neutralise heavy artillery, IS' GFs would commonly engage the SDF fighters in Raqqa from close distances, sometimes by using the tunnel networks to get closer. The GFs were also nimble with their light artillery, especially in the southeastern suburbs of Raqqa City.[167] IS mortar crews would fire and then rapidly relocate to avoid the SDF detection, who would usually send the coordinates to the Coalition for close air-support. Due to their constant mobility, the GFs outflanked the SDF in both Raqqa City and Raqqa Governorate on multiple occasions.[168] Finally, as in the urban battles of Mosul, Ramadi and Sirte, the manoeuvrability of the GFs/SGFs was enhanced by an extensive network of inter-building and intra-building holes and tunnels. As in the other cases,[169] the tunnels of Raqqa varied. Some were just a basic passage. Others were quite sophisticated. As in the words of Colonel Ryan Dillion, the OIR spokesperson:

> You'll see a concrete-reinforced tunnel system that runs underneath the children's hospital [in Raqqa City]. The entrance to the tunnel is well-constructed, with handrails and stairs. And then, as you get down into the tunnel, you'll see here, where this is the main tunnel, and then it branches off into less elaborate dirt tunnels that lead to houses and shops nearby.[170]

'ISIS has cots and cooking stoves and has also stockpiled weapons and ammunition in these structures', he explained. This meant that the tunnel network did not just enhance the GFs and SGFs manoeuvres, but they were also used as operational planning headquarters, logistical support bases, command-and-control centres and weapon depots.

To avoid further reduction of their already limited manpower, IS GFs used remote-controlled recoilless rifles, sniper rifles and heavy machine guns in Raqqa and Deir Ezzor Governorates, at least. IS workshops in the Raqqa and Deir Ezzor Provinces modified weapons so that the GFs could aim and fire them remotely. These modifications were undertaken using civilian videogame controllers and a TV screen linked to a camera attached to the weapon.[171] IS mortar crews has also used smartphones, Google Earth and other civilian applications and gadgets to guide their first ranging shot (in addition to the slower 'bracketing',[172] 'creeping fire'[173] and other traditional techniques). This usually involves two or three IS units: the firing crew, a ground-observer(s) and/or an affiliate(s) connected to Google Earth or similar applications. After the first ranging shot, the observer or the affiliate (using information given by the observer on the ground) would drop a pin on the application's map. The pin would appear on the smartphone version of the application held by the IS mortar crew, who would then be able to fire more accurately on the target.[174]

UCAVS/Drones

IS had also used drones extensively in Raqqa. In the first month of the battle of Raqqa City (June 2017), SDF fighters suffered from two drones attacks every day, at minimum.[175] In some days, these attacks exceeded fifteen. As in Mosul, IS drones in the Raqqa Governorate conducted multiple missions. Seven types could be identified: intelligence, surveillance and reconnaissance (ISR); messaging; identifying targets; propaganda; coordinating attacks; area-denial; and direct attacks. For the latter missions, IS units innovated and modified commercial drones to enable them to drop small bombs simultaneously. As a result, IS drones would be rotating attacks on the very same position for hours. In some districts in Raqqa City, this tactic denied areas to SDF snipers, such as exposed positions on rooftops.[176] IS direct drone attacks targeted entrenched positions, mobile SDF units and fixed targets in Raqqa.

Armoured units

As highlighted above, ISIS/IS military formations preferred unconventional tactics in the battlefields, whether terrorism or guerrilla warfare tactics. This is clearly reflected in the organisation's newsletter, audio-visual releases, statements and military metrics. Still, the organisation managed to capture hundreds of armoured vehicles. Perhaps more dangerously, it also knew how to use them effectively. This contradicts the common perception that the organisation is incapable of handling conventional armoured vehicles.[177]

ISIS/IS in Syria had tanks since its inception in April 2013. This was due to the capture of a part of the JN's arsenal. As early as January 2014, the organisation released a video of a column of armoured vehicles (included main-battle tanks, armoured self-propelled guns and armoured personnel carriers) in Northern Aleppo. This column included at least one modified T-72 tank. IS recruited tank crews and established specialised units which operated armoured vehicles, such as the Affan Division, al-Jarrah Division and al-Awam Division.[178] These are combined arms, company- or battalion-sized formations. They are neither division-sized nor exclusively armoured-formations in the conventional sense. Some of these were operating under the OIR airstrikes after September 2014, partly because of logistical hurdles to rapid coordination between ground-intelligence and close air-support.

Intelligence and transfers of knowhows

As mentioned earlier, ISIS/IS invested in building intelligence, surveillance and reconnaissance (ISR) capacities within their military and security structures. These investments were stimulated by lessons from the Iraqi experiences, as well as other factors (including the background of some of the IS leaders as intelligence officers, both Iraqi and non-Iraqi). In Syria, the organisation would commonly send small units to its enemies' territories to map out and collect intelligence on power-structures and bases, as well as on the manpower, the firepower, armament, training and daily routines of its enemies.[179] These units would also conduct close reconnaissance of targeted positions, attempts to infiltrate other organisations, use false-flag tactics and send/deploy spies and recruiters.[180] Although these capacities are not unique among ANSAs, IS excelled in transferring the knowhows of both military tactical innovations and ISR skills in-between its units and 'provinces'. In an interview with the *Rumiyah* English-language magazine of IS, an unnamed IS military commander of Raqqa was asked about the effect of the battle of Mosul on the battle of Raqqa. He said that the new tactics and experiences in Mosul have been passed on to all IS 'provinces' so they 'could benefit from them'.[181] This was clear in the case of the SVBIED designs and tactics. The designs that were used in Raqqa (and Sirte) were very similar to those used in Mosul, including the camouflaged SUV-based SVBIED designs.

After Raqqa: the IS insurgency in Syria

IS territorial losses and military defeats continued after losing the Raqqa Governorate. In the nearby Deir Ezzor Governorate, ISIS/IS has been fighting almost all state and nonstate armed actors operating in that governorate. Between April 2014 and January 2020, the organisation had launched or fought back fourteen military campaigns in Deir Ezzor. It combatted units from at least ten ASAs and twelve ANSAs or coalitions of ANSAs.[182] Nine of these campaigns were offensive in nature; five were defensive. Unlike the outcome in Raqqa Governorate, IS failed to control the whole of Deir Ezzor Governorate despite controlling major parts of it including parts of its provincial capital, Deir Ezzor City.

In 2019, IS made a last territorial military-stance involving thousands of its fighters in the Hajin Pocket in southeast Deir Ezzor. The pocket is a string of small towns and villages situated along the Euphrates river, before it crosses into Iraq. It contained thousands of IS fighters and thousands of their families. The pocket was populated by waves of retreating IS fighters, including a large number of IS fighters from the convoy that left Raqqa City in a compromise with the SDF in October 2017. Generally, the organisation kept on retreating from the town of Hajin in the northwest of the pocket to the town of al-Baghuz al-Fawqani (Upper Baghuz) in the southeast. IS territorial control in Syria was reduced from about half of the country's territories in 2015 to a massive tent city in and around the town of Upper Baghuz in 2019. After one month and two weeks of intense fighting and bombardment, stalemates and surrenders, the town was cleared of thousands of IS fighters on 23 March 2019. During the fighting, the organisation was still capable of innovating SVBIED designs and tactics as well as shifting ways of warfare. A week after IS was defeated, the SDF discovered an IS SVBIED workshop under a tent, near the Baghuz 'tent city' on the northern banks of the Euphrates.[183] IS was continuing its production of SVBIEDs and developing different designs, while being on the brink of territorial collapse in Syria.

Overall, the US played a major role, if not *the* major role, in combatting IS in Syria. During the battle for Raqqa City, from June to October 2017, US aircraft dropped just under 20,000 total munitions.[184] A US artillery battalion 'fired more rounds in five months in Raqqa than any other Marine artillery battalion, or any Marine or Army battalion, since the Vietnam war', said Army Sergeant Major John Wayne Troxell, a senior adviser to the chairman of the Joint Chiefs of Staff.[185] On the night of 26–7 October 2019, the US conducted a military operation in Berisha village in Syria's Idlib Governorate. The operation was codenamed Kayla Mueller[186] and its outcome was the decapitation of IS. 'The United States brought the world's number one terrorist leader to justice. Abu Bakr al-Baghdadi is dead', President Trump announced on 27 October 2019. As expected, the decapitation did not end the group in Syria or elsewhere. On 22 December 2019, IS began its 'revenge' operations for the killing of al-Baghdadi (and the former IS spokesman Abu al-Hassan al-Muhajir).[187] In the first week of the 'revenge', IS claimed responsibility for 106 attacks, 50 of which were in Syria and 34 of which were Syria-based attacks in Raqqa and Deir Ezzor Governorates.[188] In the first two weeks of 2020, the organisation claimed over 15 attacks in Syria. These attacks combined IED, SVBIEDs, GFs tactics as well as close-quarter assassinations. All the attacks targeted the SDF and the regime's forces in Raqqa and Deir Ezzor Governorates, with only two attacks executed in Homs Governorate.[189] Between March 2019, the month of IS defeat in Baghuz, and March 2020, the organisation claimed responsibility for 977 attacks in Syria. The majority of them were in Raqqa (153 attacks) and in Deir Ezzor (581 attacks). The rest of the attacks were perpetrated in Hasaka, Homs, Aleppo, Damascus, Rif Dimashq, Deraa and Quneitra. IS also launched an assassination campaign in Deir Ezzor, claiming 34 operations targeting the SDF and the regime's commanders and affiliates between January and March 2020.[190] On 31 March 2020, IS' Levant Province released a video entitled 'The Epic of Attrition 2'. The release showed some of the IS attacks throughout 2019 in al-Badiya/Homs desert (between western Deir Ezzor and eastern Homs governorates). The organisation still held a significant guerrilla arsenal in al-Badiya, including technicals fitted with heavy machines, unguided rocket launchers and ATGMs. This is in addition to several platoon-sized guerrilla formations.

Overall, IS proved highly resilient and adaptive in Syria, like in Iraq and in some of the other countries. Although many of the operations launched were unsuccessful; many of the tactics employed were broadly effective. The combinations of IEDs, SVBIEDs, snipers and armed drones surprised the organisation's enemies. Usually, IS knew how to capitalise on the element of surprise. These combinations broke momentum, slowed down the advances of attackers, harassed the SDF and other opponents, and at some point, wrought havoc (especially with the barrages of SVBIEDs or 'shelling by blowing-up' tactic). The GFs of the organisations constantly showed tactical aggression and implemented tenacious counteroffensives. At different phases and timeframes in the battles of Raqqa and Deir Ezzor Governorates, the GFs showed autonomous initiatives and innovative improvisation, especially while using tunnel warfare to stage attacks behind the SDF lines. IS unit cohesion and leadership roles were tested to their limits in Syria. Due to the limited manpower, weak air-defence systems and other disadvantaging factors; IS command-and-control at the operational-level and sometimes at the strategic-level,[191] would usually collapse during defensive battles. This led to the de-centralisation of small, cohesive units who were tactically capable of holding on to positions or of raising the costs of eliminating them and liberating these held positions. The 'hell roundabout' in Raqqa city centre serves an example.

At an operational level, the organisation had shown capacity to rapidly build up a military force from close to nothing. The five-pillar 'iALLTR' *modus operandi* aiming to fashion a force was relatively successful in Syria. It will be repeated, in part or whole, in Libya, Egypt and elsewhere with mixed results. IS three-phase operational-level *modus operandi* 'SCCLC' implemented during the occupation campaigns of cities and governorates was only successful in the case of Raqqa City. Comparatively, it was more successful when it was executed in Iraqi cities and towns. Still, SCCLC will be executed elsewhere despite its overwhelming operational failure in Syria.[192]

In terms of the future, IS and its predecessors in Syria had also demonstrated a capacity to rebuild their military capabilities after being on the brink of collapse. Like in Iraq, the organisation mastered strategic shifts between conventional, guerrilla and terrorism ways of warfare to avoid annihilation in Syria. The above review and analysis may provide some forecasting insights into how IS (and similar armed organisations) will fight in Syria in the future, given that the 'know-how' is available and the monopoly over the means of violence has been broken by ANSAs.[193] On a macro-level, the future of IS in Syria is likely to be affected by the destruction of major parts of Raqqa, Deir Ezzor and many other towns and cities, the endurance of the Syrian civil war, the lack of a credible or a sustainable conflict resolution mechanism(s) and the overall brutal environment that regime's policies and its allies have created. The organisation or variants of it are more likely to endure in such an environment, and perhaps rise again from its ashes in Syria.

Notes

1. Interview with Abu Sa'id, a fighter from *Ahrar al-Sham* Quoted in: Hamzeh al-Mustafa, 'The Islamic State in Syria: Origins and the Environment', [in Arabic] in: Azmi Bishara (editor), *The State Organisation (Acronym ISIS): Foundation, Discourse and Practice*, [in Arabic] vol. 2 (Doha: the Arab Centre for Research and Policy Studies), p. 301.
2. Alex Horton, 'ISIS fighters booby-trapped corpses, toys and a teddy bear in besieged Raqqa', *The Washington Post*, 18/10/2017.
3. Shawn Snow, 'These Marines in Syria fired more artillery than any battalion since Vietnam', *Marine Corps Time*, 6/2/2018.
4. Al-Manara Al-Bayda Corporation for Media Production, 'Video: Announcing the Formation of the Suppor Front in the Levant to Fight Bashar al-Assad', [in Arabic] *Dailymotion*, 12/ 2011, Accessed on 25/3/2018, at: https://bit.ly/2QFP6Ra
5. According to a Damascus Criminal Court case charging him with terrorism, the real name of Abu Muhammad al-Julani is Muhammad Hussein al-Shar'. See: Mohamed Manar Hamijou, 'As a Result of Personal Allegations from Victims, in Addition to the Public Prosecution . . . Death Sentences in Absentia for 'Al-Julani', 'Al-Buwaydani', 'Al-Shameer' and Dozens of Terrorists', [in Arabic] *Al Watan*, 11/12/2018, Accessed on 12/1/2019, at: http://alwatan.sy/archives/178725

   However, in a public meeting with tribal leaders from Jabal al-Zawiya, he identified himself as 'Ahmed Hussein al-Shar''. Born in 1980 or 1981, he joined the Iraqi insurgency during his second year of college (studying Media at the University of Damascus). He was arrested in Iraq and released on 11 March 2011. See for example: 'Al-Julani is a Relative of Faruq al-Shar' [. . .] He Studied Jurisprudence at the Hands of a Damascene Scholar in Mezze', [in Arabic] *Enab Baladi*, 28/7/2016, Accessed on 14/4/2020, at: https://bit.ly/34ypU4V
6. Any affiliation with al-Qaida and the ISI was consistently denied by the JN leadership until April 2013.
7. Interview with Moussa Al-Ghannami, in: Ja'far al-Wardi, 'The dispute between al-Nusra and ISIS', [in Arabic], *Kawalis al-Thawra* - Episode 42, *Dar Eman TV*, 21/5/2016, accessed on 25/3/2017, at: https://bit.ly/2XyabkN
8. Rania Abouzeid, 'The Jihad Next Door: The Syrian Roots of Iraq's Newest Civil War', *Politico*, 23/6/2014.
9. Salih Hamah (also known as Abu Muhmmad al-Hamawi, and – on Twitter – '*Us al-Sira' fi al-Sham* [Core of the Conflict in the Levant]) was an ISI commander and a co-founder of al-Nusra Front (JN), before being expelled from the organisation and becoming a major critic of the JN, the ISI and their successor organisations. Saleh Al-Hamawi (real name is 'Atif al-Tirkawi is a dentist and a known activist in the humanitarian assistance field. See: 'Jabhat Al-Nusra Expels the Leader '*Us al-Sira' fi al-Sham* Permanently',

[in Arabic], *Arabi 21*, 15/7/2015, Accessed on 14/4/2020, at: https://bit. ly/2RxjhdK; 'Al-Nusra Expels One of its Founders, Social Media Accounts and Websites Confuse him with Saleh al-Hamawi', [in Arabic] *Zaman Al Wasl*, 16/7/2015, Accessed on 14/4/2020, at: https://bit.ly/34DRAoT . Many English- and Arabic-language sources conflate the two persons.

10. Iyad Tubasi, a Jordanian-Palestinian ISI commander, is the brother-in-law of Abu Musab al-Zarqawi, the founder of AQI and other predecessors of IS.

11. Mustafa al-Salih, a Jordanian ISI commander, was in charge of recruitment and logistical support.

12. Maysara al-Juburi, an Iraqi ISI commander who will become the deputy leader of *Jabhat al-Nusra* (JN), and later on an archenemy of ISIS. He was already in dispute with ISI's leaders and he has settled in Syria since 2009. He opened a grocery shop in Deir Ezzor. Hence, ISI/ISIS/IS members would mockingly refer to him as 'Abu Marir al-Baqqal' (father of bitterness, the grocer).

13. The LIFG believed that allying with Algerian Jihadists (especially the *Groupe Islamique Armé* or GIA) gave them a strategic depth. If the Algerian regime was toppled or if parts of western Algeria were controlled by the GIA, the LIFG would have a contiguous border, across which it could launch attacks into Libya. The 'leap' (*al-wathba*) was a term used by the LIFG members – interviewed for an earlier study – to describe the move of trained fighters from Afghanistan to Algeria, a country that shares 610 miles of borders with Libya. Although the 'leap' seemed strategically sound on paper, it ultimately proved to be a disaster for the LIFG. The ISI 'leap' into Syria was much more successful. See: Omar Ashour, 'Post-Jihadism: Libya and the Global Transformations of Armed Islamist Movements', *Terrorism and Political Violence*, vol 13, no. 3, p. 382.

14. See for example: Craig Whiteside, 'New Masters of Revolutionary Warfare: The Islamic State Movement (2002–2016)', *Perspective on Terrorism*, vol. 10, no. 4, August 2016, accessed on 25/3/2020, at: https://bit.ly/3cib7xW

15. The *Sham* is equivalent to the narrowest possible 'Levant' in historical sense: Syria, Jordan, Israel, Palestine and Lebanon. In other words, it is equivalent to 'Greater Syria'. For the audio statement, see: al-Wardi, *Kawalis al-Thawra -* Episode 42.

16. Ibid.

17. Ibid.

18. For more details see: Mohamed Abu Rumman and Hasan Abu Haniya, *The Islamic State: The Sunni Crisis and the Struggle over Global Jihadism*, [in Arabic] (Amman: Friedrich-Ebert-Stiftung, 2015); Hassan Hassan, 'Two Houses Divided: How Conflict in Syria Shaped the Future of Jihadism', *CTC Sentinel*, vol. 11, no. 9 (October 2018), Accessed on 14/4/2019, at: https://bit. ly/34AbkK4

19. Muhammad Bahhaya (Abu Khalid al-Suri and Abu 'Umayr al-Shami) was killed in Aleppo's Operational Room by a squad of ISIS' SGF. See: Ahmad Aba Zeid, 'Abu Khaled Al-Suri: The First Generation in the Face

of the Last Deviation', [in Arabic] *Zaman Al Wasl*, 26/2/2015, Accessed on 14/4/2019, at: https://bit.ly/3eecJpv; 'Jihadist Leader Abu Khaled Al-Suri was Killed in an Attack in Aleppo', [in Arabic] *BBC*, 24/2/2014, Accessed on 14/4/2019, at: https://bbc.in/34wZ90K; 'Pictures . . . Abu Khaled Al-Suri from Birth to Death in Aleppo', [in Arabic] *Taht Al Mijhar*, 24/2/2014, Accessed on 14/4/2019, at: https://bit.ly/2Vpg5lK

20. Kareem Shaheen, 'ISIS "Controls 50% of Syria" After Seizing Historic City of Palmyra', *The Guardian*, 21/5/2015, Accessed on 14/4/2017, at: https://bit.ly/2XxZcI4

    For a map of IS territorial losses and gains see: 'Islamic State Losses Between January 2015 and 3 April 2017', *Business Wire*, 2017, Accessed on 14/4/2018, at: https://bwnews.pr/34B0iED

21. The support of the JN and the coalitions it led (*Jabhat Fatih al-Sham* and *Hay'at Tahrir al-Sham*) considerably fluctuated between 2012 and 2020.

22. Interview with Abu Sa'id, a fighter from *Ahrar al-Sham* Quoted in: Hamzeh al-Mustafa, 'The Islamic State in Syria: Origins and the Environment', [in Arabic] in: Azmi Bishara (editor) *The State Organisation (Acronym ISIS): Foundation, Discourse and Practice*, [in Arabic] vol. 2 (Doha: the Arab Centre for Research and Policy Studies), p. 301.

23. More details on the January 2014 battle of Raqqa City are found below.

24. The ten 'provinces' are al-Barakah (al-Hasaka), al-Khayr (Deir Ezzor), al-Raqqa, Homs/al-Badiya, Halab, Idlib, Hama, Damascus, al-Sahel (Latakia and Tartous) and Horan (or Army of Khaled Ibn al-Walid). IS also claimed al-Furat (Euphrates) Province, which claimed territories in both Syria and Iraq. In July 2016, IS issued a video entitled *Sarh al-Khilafah* (Structure of the Caliphate) in which it proclaimed only seven provinces in Syria, plus al-Furat. It was an updated proclamation, reflecting the losses of al-Sahel and Idlib; 'Horan Province' was still in the making. By July 2018, IS had reorganised and merged all of its 'provinces' in response to territorial losses and series of tactical and operational defeats.

25. On a macro-level, this book argues that the repressive, corrupt, anti-liberal, xenophobic, sectarian and misogynist policies of Gaddafi, Assad, Mubarak and al-Sisi regimes as well as al-Maliki's government in Iraq had created hospitable environments for IS and like-minded organisation to exist, expand and endure. For an analysis of the macro-level see for example: Azmi Bishara, *The State Organisation (Acronymed ISIS) A General Framework and a Contribution to Help Understand the Phenomenon*, [in Arabic] vol. 1 (Doha: the Arab Centre for Research and Policy Studies), 2018.

26. This policy of selective 'collusion and collision' continued during the Syrian civil war.

27. In 21 interviews conducted by the author for this chapter with Syrian commanders, fighters and soldiers from different sides and organisations (that all fought ISIS/IS in Syria), one of the very few conclusions that all interviewees agree on is the strong intelligence capacities of ISI/ISIS/IS.

28. Non-locals include both non-Syrians and fighters from outside of the governorates in which they are deployed. Both types are usually unfamiliar with the geography/topography, tribal, clan and social power maps.

29. See for example Charles Lister, *The Syrian Jihad* (London: Hurst, 2015), pp. 37–39.

30. Christoph Reuter, 'Secret Files Reveal the Structure of Islamic State', *Spiegel International*, 18/4/2015, Accessed on 19/4/2019, at: https://bit.ly/2RAR7i4

31. Ibid.

32. The Army of Islam was a Sunni-Salafist armed Islamist organisation that became one of the most militarily effective Syrian rebel organisations in southern Syria. It operated mainly in the south (with few units in northern and central Syria) and had its stronghold in the strategic eastern Ghouta area close to the capital of Damascus, before losing it to the regime's forces.

33. Former Army of Islam Commander, Interview by author, Istanbul, 7 November 2016.

34. Ibid.

35. Ibid. Alloush did not know the background of al-Julani and what he represented at the time. Under Alloush's command, the Army of Islam launched a war against ISIS in southern Syria as of late 2013.

36. David Ignatius, 'Al-Qaeda Affiliate Playing Larger Role in Syria Rebellion', *Washington Post*, 30/11/2012, Accessed on 14/4/2016, at: https://wapo.st/2VpN2hH

    The 10,000 figure is probably an overestimate. Other FSA and armed opposition sources interviewed by the author estimate the manpower to be less than half of this figure at that time. Surveying the manpower of JN units in each 'area of operation' or 'area of influence', the average estimate was about 4,000 fighters within that timeframe.

37. One FSA commander claimed to the author that about 90 per cent of JN's foreign fighters joined ISIS after the al-Baghdadi's declaration in April 2013.

38. Jim Sciutto et al., 'ISIS Can "Muster" Between 20,000 and 31,500 Fighters, CIA says', *CNN*, 12 September 2014, Accessed on 14/4/2016, at: https://cnn.it/34x1Jns

39. 'Around 585,000 People Have Been Killed Since the Start of the Syrian Revolution, Calling for Freedom and Democracy', *The Syrian Observatory for Human Rights*, 4/1/2020, Accessed on 14/4/2020, at: https://bit.ly/2xvxTDw

40. The Syrian Democratic Forces (SDF), member-organisations of it such as the YPG and YPJ, and related civilian institutions have a consistent tendency to inflate the numbers of ISIS/IS fighters and other jihadists. Given the bias, bitterness and bloodshed engendered by the Syrian civil war the estimates should be taken with caution.

41. See for example: Aaron Y. Zelin, 'Wilayat al-Hawl: "Remaining" and Incubating the Next Islamic State Generation', *Policy Notes*, The Washington

Institute for Near East Policy, October 2019, Accessed on 14/4/2020, at: https://bit.ly/2RCZARN

42. See for example the interview with Major-General Alex Grynkewich, the deputy commander of the American-led coalition to defeat IS. He estimated that al-Hawl Camp alone has '20,000 suspected 'hardcore IS' members'. See: Lara Seligman, 'In Overflowing Syrian Refugee Camps, Extremism Takes Root', *Foreign Policy*, 29 July 2019, Accessed on 14/4/2020, at: https://bit.ly/2V7M76B

43. For details, see the section entitled 'Striking from the Ashes: Overview of the IS Military Build-Up in Iraq' in Chapter 2 of this book.

44. Lucy Draper, 'ISIS Controls over 50% of Syria After Taking Palmyra', *Newsweek*, 21/5/2015, Accessed on 14/4/2020, at: https://bit.ly/2VrrTnp

45. This is shown in the IS photographic reports and propaganda video releases by *al-Khayr* province. The release series are entitled 'At the Doors of Epics [Battles]'.

46. The conservative estimate is based on pictures and videos of weapon published by the media offices of seven of ISIS/IS' Syrian Provinces, in addition to IS/ISIS publications between April 2013 and December 2014. By the end of 2015, the organisation had looted and operated around 200 tanks in Syria.

47. See for example the tanks and 130mm artillery pieces looted only from the 93rd Bridge in 'Ain 'Iss (Raqqa Governorate): 'Video Leaked from Brigade 93, Raqqa, after ISIS Took Control of it', *Step News*, 7/8/2014, Accessed on 25/3/2020, at: https://bit.ly/3ds0JEZ

48. The majority of the infantry fighting vehicles recorded are Soviet-made BMP-1 (amphibious tracked). Some of which were modified and upgraded with a DShK heavy machine gun (12.7 mm).

49. A. W. Former FSA Commander in Raqqa, Interview conducted by author, Istanbul, 17 September 2017.

50. Ibid.

51. Ibid.

52. Many of the tanks and the artillery pieces were taken from the regime's 93rd armoured brigade, the 21st artillery regiment and 17th special forces division bases during July and August 2014. Other weapons were taken from al-Tabqa airbase in the same large-scale 'Northeastern Offensive' of summer 2014 in which IS forces attacked the military installations of the regime in northeastern Syria (mainly Raqqa, al-Hasaka and Aleppo). See for example: 'Jihadists Capture Key Base from Syrian Army', *The Daily Star Lebanon*, 8/8/2014, Accessed on 14/4/2020, at: https://bit.ly/2JZJw8C; Abdullah Suleiman Ali, 'Showdown Begins between Syrian Army, Islamic State', *Al-Monitor*, 25/7/2014, Accessed on 14/4/2020, at: https://bit.ly/2xspPDI

53. Ibid.

54. Many of these rockets, and also ATGMs, heavy machine guns and other weapons, were looted from the 121st Regiment (al-Milbiyah Regiment) in al-Hasaka Governorate in July 2014.

55. These were looted from al-Tabqa military airbase in Raqqa in August 2014. Jeremy Bender, 'ISIS Just Looted Advanced Weaponry from a Crucial Assad Regime Air Base in Syria', *Business Insider*, 25/8/2014, Accessed on 14/4/2018, at: https://bit.ly/3a88cGy

56. Ibid.

57. NATO reporting name of the Soviet-made S-125 *Neva/Pechora*. Abu Qutaybah. Former FSA Commander, Interview by author, Gaziantep, 12 November 2018.

58. These were looted from al-Tabqa military airbase in Raqqa. The AA-2 (NATO's designation code for Soviet-made Vympel K-13 air-to-air missile) is commonly confused with the American-made AIM-9 Sidewinder, due to similarities in appearance. IS managed to use air-to-air missiles in ground fighting mainly as IEDs or variants of them.

59. Ahmed Al-Arabi, 'ISIS Conducts Airstrikes in Raqqa', [in Arabic] *Aljazeera*, 28/10/2014, Accessed on 14/4/2020, at: https://bit.ly/2yTGu3i

60. Ibid.

61. 'Weapons of the Islamic State: A Three-Year Investigation in Iraq and Syria', *Conflict Armament Research* (12/2017), pp. 183–4; 189, Accessed on 14/4/2020, at: https://bit.ly/2K5I0lb

62. Ibid., p. 184.

63. Former Syrian Army Officer, Interview by author, Beirut, 6 December 2017; Former Commander in *Liwa al-Haqq* (Truth Brigade), Interview by author, Istanbul, 7 October 2016; 'Including Missile Launchers and Artillery . . . The State' Parades the 'Spoils' of the 121st Regiment', *Zaman Al-Wasl*, 28/7/2014, Accessed on 14/4/2020, at: https://bit.ly/2xvLzP5

64. For the background of dissolving all non-IED tactical units in the ISI, including sniping units and guerrilla formations and focusing all ISI's resources on building IED-tactical units, see the section entitled 'Striking from the Ashes: Overview of the IS Military Build-Up in Iraq' in Chapter 1.

65. Hugo Kamaan, 'Islamic State SVBIED Development since 2014', paper presented at the annual conference of the Strategic Studies Unit entitled 'Militias and Armies: Developments of Combat Capacities of Armed Non-State and State Actors', Arab Centre for Research and Policy Studies (ACRPS), Doha, 24/2/2020, Accessed on 14/4/2020, at: https://bit.ly/2RAQh4J

66. Former Syrian Army Officer, Interview by author, Beirut, 6 December 2017; Abu Qutaybah, Former FSA Commander, Interview by author, Gaziantep, 12 November 2018.

67. Bill Roggio and Caleb Weiss, 'More Jihadist Training Camps Identified in Iraq and Syria', *FDD's Long War Journal*, 23/11/2014, Accessed on 14/4/2019, at: https://bit.ly/3a4kwHL

68. Abu Qutaybah, Former FSA Commander, Interview by author, Gaziantep, November 2018; Former Commander in the FSA and *al-Jabha al-Shamiya* (Levantine Front), Interview by author, Gaziantep, November 2018.

69. In addition to the JN and ISIS, there are other smaller affiliates, loyalists and (shifting) tactical allies of ISIS/IS in Syria ranging from *Jund al-Aqsa*

(Soldier of the Aqsa Mosque) in the north to Khaled ibn al-Walid Army in the south.

70. Azmi Bishara (editor), *The State Organisation (Acronym ISIS): Foundation, Discourse and Practice,* [in Arabic] vol. 2 (Doha: the Arab Centre for Research and Policy Studies), 2018, p. 299.

71. These included FSA and *Ahrar al-Sham* factions, in addition to the JN. The organisation also recruited and bought arms from Assad's soldiers based in Raqqa.

72. For details, see the following chapter.

73. Kidnapped figures included Abdullah al-Khalil, the head of the Local Council of Raqqa Governorate (established via elections after liberation from the regime), Firas al-Hajj, a leading civil activist, Father Paolo Dall'Oglio, a Jesuit Priest who criticised the brutal policies of the Assad regime, among others. These victims did not have any military value. They just represented potential leaders who may have been able to mobilise civilians against ISIS.

74. For details about the infiltrator and an eyewitness description of the operations see: Ibrahim Darwich, 'The Story of the Fall of Raqqa in the Hands of Jihadists . . . Betrayals, Executions and Cruel Practices', [in Arabic] *Al Quds Al Arabi,* 16/6/2015, Accessed on 14/4/2020, at: https://bit.ly/2XyNrBi

75. Al-Hadrami was a JN commander who chose the ISIS side in the April 2013 split. He became the *emir* of ISIS in Raqqa City. Due to many disagreements and clashes with the ISIS *emir* of Raqqa Governorate, Abu Luqman (Ali Musa al-Shawwah), al-Hadrami defected to JN and became its commander in Raqqa City.

76. Those organisations were *Ahrar al-Sham, Jabhat al-Nusra* (JN), and three factions operating under the FSA umbrella in Northeastern Syria (*Ahfad al-Rasul* [Descendants of the Messenger] Brigades, *Thuwar al-Raqqa* [Revolutionaries of Raqqa] Brigade, and *al-Muntasir Billah* [The Victor with God's Help] Brigade). ISIS executed al-Hadrami himself in January 2014, after winning the battle of Raqqa City.

77. See Chapter 1 on Iraq and Chapter 3 on Libya.

78. Interview with Abu Sa'id, a fighter from *Ahrar al-Sham* Quoted in: Hamzeh al-Mustafa, 'The Islamic State in Syria: Origins and the Environment', [in Arabic] in: Azmi Bishara (editor) *The State Organisation (Acronym ISIS): Foundation, Discourse and Practice,* [in Arabic] vol. 2, the Arab Centre for Research and Policy Studies, p. 301.

79. Ibid. p. 301; Saad S. Former FSA fighter in Raqqa and Deir Ezzor, Interview by author, Gaziantep, November 2018.

80. Ibid; Abu al-Harith, Former Army of Islam – North Commander, Interview by author, Gaziantep, November 2018.

81. Ibid. Wael Issam, 'Al-Quds al-Arabi Narrates the Reasons for the Sudden Fall of Raqqa in the Hands of the Islamic State . . . the Relationship Between Jabhat al-Nusra and ISIS in the City . . . from Alliance to War', [in Arabic] *Al Quds Al Arabi,* 30/1/2014.

82. A. W. Former FSA Commander in Raqqa, Interview by author, Istanbul, 9 October 2016. See also: Wael Issam, 'The Leader of al-Nusra Front for "Al-Quds Al-Arabi": ISIS Fighters were Calling Us on the Radio Saying "Ahrar al-Sham Withdrew and Left You"', *Al Quds Al Arabi*, 31/1/2014, Accessed on 14/4/2020, at: https://bit.ly/3ejZKay
83. Ibid.
84. Issam, 'Al-Quds Al-Arabi Narrates the Reasons for the Sudden Fall of Raqqa'.
85. The sources are unclear on how they passed through. Most interviewees on that subject argue that this was a conspiracy between the JN and ISIS (almost all of the interviewees fought against both organisations). The conspiracy argument does not explain the fighting that took place between the JN and ISIS in the eastern neighbourhoods of Raqqa City and al-Tabqa afterwards, as well as the executions of the JN prisoners by ISIS.
86. At least two SGFs managed to infiltrate the *Ahrar*'s positions in the late night of the 10th and early hours of 11th January.
87. Interview with Abu Sa'id, a fighter from *Ahrar al-Sham* Quoted in: Hamzeh al-Mustafa, 'The Islamic State in Syria: Origins and the Environment', [in Arabic] in: Azmi Bishara (editor) *The State Organisation (Acronym ISIS): Foundation, Discourse and Practice*, [in Arabic] vol. 2, the Arab Centre for Research and Policy Studies, p. 301.
88. Interview with Abu Salih, a doctor in *Ahrar al-Sham*'s makeshift who eyewitnessed the battle, Interview by the research team on 25/2/2015, Quoted in: Hamzeh al-Mustafa, 'The Islamic State in Syria: Origins and the Environment', [in Arabic] in: Azmi Bishara (editor) *The State Organisation (Acronym ISIS): Foundation, Discourse and Practice*, [in Arabic] vol. 2, the Arab Centre for Research and Policy Studies, p. 300.
89. Abu Rajab, Former FSA Fighter who fought ISIS/IS in Idlib, Aleppo and Raqqa, Interview by author, Istanbul, 9 October 2016; See also: Wael Issam, 'Al-Quds Al-Arabi Narrates the Reasons for the Sudden Fall of Raqqa in the Hands of the Islamic State . . . the Relationship Between Jabhat al-Nusra and ISIS in the City . . . from Alliance to War', [in Arabic] *Al Quds Al Arabi*, 30/1/2014; 'The Sudden Fall of Raqqa', [in Arabic] *Safahat Souriyya*, 1/2/2014, Accessed on 14/4/2020, at: https://bit.ly/3a3EC57
90. Some anti-ISIS resistance was still occurring in Raqqa as late as March 2014. For example, on 2 February 2014, five wounded ISIS fighters were assassinated in the Raqqa National Hospital. On 3 February 2014, the FSA's Revolutionaries of Raqqa (*Thuwar al-Raqqa*) Brigade launched a coordinated attack on four ISIS checkpoints in the city. Abu Mansur. Former FSA Fighter (Revolutionaries of Raqqa Brigade), Interview by author, Istanbul, November 2018.
91. Ibid.
92. Ibid. See also: Al-Majeed al-Olwani, 'After Withdrawals and 'Betrayals' . . . Raqqa Rebels Regain 50% of the City', [in Arabic] *Orient Net*, 13/1/2014, Accessed on 14/4/2018, at: https://bit.ly/34KMrM5

93. The town is considered to be an ISIS stronghold in the Raqqa Governorate, prior to 2014.
94. The 121st Artillery Regiment, a regiment within the 17th Division, was also under attack as of 24 July 2014. However, it is located in nearby Milbiyah in al-Hasaka Governorate outside (but near) the Raqqa Governorate.
95. Former Syrian Army Officer, Interview by author, Beirut, 6 December 2017.
96. IS media covered parts of the battle in an English-language propaganda documentary entitled 'Flames of War: The Fighting has Just Begun'. It was released on 17 September 2014 by al-Hayat (The Life) Media Centre. In that documentary, an IS commander claimed that 'only tens of IS fighters' attacked the 17th Division. However, four FSA commanders who fought in Raqqa Governorate and were interviewed by the author estimated the number to be about 300, divided on six formations. See also: 'Division 17th Soldiers Dug their Graves with their Own Hands Before they were Executed in "Flames of War"', [in Arabic] Zaman Al Wasl, 22/6/2014, accessed on 14/4/2020, at: https://bit.ly/2Xx4kw8
97. 'ISIS Announces Control over the 17th Division', [in Arabic] Zaman Al Wasl, 25/7/2014, Accessed on 14/4/2020, at: https://bit.ly/2XB0LF1
98. Former Syrian Army Officer, Interview by author, Beirut, 6 December 2017.
99. Ibid.
100. Ibid. Wasim Nasr, 'Details of the Open Confrontation Between the "Islamic State" and the Syrian Army', [in Arabic] France 24, 28/4/2014, accessed on 14/4/2020, at: https://bit.ly/3a6gSgK; 'By the Video: The Islamic State Takes Control Over the 17th Division in the Countryside of Raqqa', [in Arabic] BBC News, 27/7/2014, Accessed on 14/4/2020, at: https://bit.ly/3b9Zd8W; 'ISIS Controls the 17th Division in Raqqa', [in Arabic] Al Jazeera, 25/7/2014, Accessed on 14/4/2020, at: https://bit.ly/3b39ns5
101. 'The "Islamic State" Storms the 17th Division of the Army in Raqqa', [in Arabic] Annahar, 25 July 2014, Accessed on 14/4/2020, at: https://bit.ly/3adPJsl
102. 'The "Islamic State" Controls the 17th Division of Raqqa', [in Arabic] Alkhaleej Online, 25/7/2014, Accessed on 14/4/2020, at: https://bit.ly/2Rz7Bam
103. The armed opposition attacks and the siege has certainly softened the resistance inside the 17th Division. IS has capitalised on that.
104. For an analysis of the Milbiyah offensive see Barak Barfi, 'The Military Doctrine of the Islamic State and the Limits of Ba'athist Influence', CTC Sentinel, vol. 9, no. 2 (February 2016), p. 19.
105. Nasr, 'Details of the Open Confrontation'.
106. Former Syrian Army Officer, Interview by author, Beirut, 6 December 2017; A. W. Former FSA Commander in Raqqa, Interview by author, Istanbul, 9 October 2016.
107. Ibid.

108. 'ISIS Controls 93rd Brigade in Raqqa's Countryside', [in Arabic] *Aljazeera*, 7/8/2014; 'Syria: ISIS controls the headquarters of 93rd Brigade', [in Arabic] *The Syrian Observatory for Human Rights*, 8/8/2014.

109. 'Jihadists Capture Key Base from Syrian Army', *The Daily Star*, 8/8/2014, Accessed on 14/4/2020, at: https://bit.ly/3eeSkp3; 'Syrian Activists: ISIS Stormed the Headquarters of the 93rd Brigade in Raqqa Causing Dozens of Dead Soldiers', [in Arabic] *CNN Arabic*, 7/8/2014, Accessed on 14/4/2020, at: https://cnn.it/2xjPSwZ; 'ISIS Controls the 93rd Brigade in Raqqa's Countryside', [in Arabic] *Al Jazeera*, 7/8/2014, Accessed on 26/3/2020, at: https://bit.ly/3a2vzRO

110. 'Four Pieces of Information about the Tabqa Military Airport in Raqqa', [in Arabic] *Enab Baladi*, 27/3/2017, Accessed on 14/4/2020, at: https://bit.ly/2RDkiB3

111. Sources close to the regime place the figure at about 1,400 officers and soldiers. See the overview in: 'Military Officials Reveal New Facts about the Fall of Tabqa Airport', [in Arabic] *Alsouria.Net*, 27/2/2016, Accessed on 14/4/2020, at: https://bit.ly/2Vuh5om; 'Four Pieces of Information about the Tabqa Military Airport in Raqqa', *Enab Baladi*.

112. A. W. Former FSA Commander in Raqqa, Interview by author, Istanbul, 9 October 2016; Former Syrian Army Officer, Interview by author, Beirut, 6 December 2017.

113. Ibid.

114. Ibid.

115. Ibid. See also: 'US Gives Syria Intelligence on Jihadists: Sources', *The Daily Lebanon Star*, 26/8/ 2014, Accessed on 14/4/2020, at: https://bit.ly/2wGuiCo

116. 'ISIS Ends the Battle for Tabqa Airport, the Regime Admits its Loss, and the Shock Silences the Regime's Supporters', [in Arabic] *Zaman Al Wasl*, 24/11/2014, Accessed on 14/4/2020, at: https://bit.ly/3b7MG6f

117. Ibid; A. W. Former FSA Commander in Raqqa, Interview by author, Istanbul, 9 October 2016; Former Syrian Army Officer, Interview by author, Beirut, 6 December 2017.

118. Sylvia Westall, 'Hundreds Dead as Islamic State seizes Syrian Air Base - Monitor', *Reuters*, 24/8/2014, Accessed on 14/4/2020, at: https://reut.rs/3a9c8qD; Wasim Nasr, 'The Repercussions of the Fall of Tabqa Military Airport under ISIS', [in Arabic] *France 24*, 29/8/2014, Accessed on 14/4/2020, at: https://bit.ly/2V5EJZu

119. A mainly Kurdish light- and motorised-infantry militia in Syria. It started as an armed wing of the 'Kurdish Supreme Committee' composed of the Democratic Union Party (PYD) and the Kurdish National Council (KNC) in 2011, although its origins and predecessors can be traced back to the 2004 Qamishli riots and earlier. Gradually, the YPG became one of the most effective military actors in northeastern Syria.

120. The armed nonstate actors fighting IS in Raqqa were: YPG, YPJ (Women's Protection Units), Revolutionaries Army, Kurdish Front,

Democratic North Brigade, The Clans Forces, The Brigade of Homs Guerrillas/Commandos, Raqqa Hawks, The Liberation Brigade, The Seljuks Brigade, The *Sanadid* (Brave) Force, Syriac Military Council, Manbij Military Council, Deir Ezzor Military Council, Elite Forces, Self-Defense Forces and armed units associated with the Civil Council of Raqqa and the Syrian Democratic Council.

121. Quentin Sommerville and Riam Dalati, 'Raqqa's Dirty Secret', *BBC News*, Accessed on 26/3/2020, at: https://bbc.in/3ae0meI; Dominic Evans and Orhan Coskun, 'Defector Says Thousands of Islamic State Fighters Left Raqqa in Secret Deal', *Reuters*, 7/12/2017, Accessed on 14/4/2020, at: https://reut.rs/2XAzm6q

122. Estimates of IS fighters in Raqqa were in the range of 4,000. Estimates of the SDF and Coalition forces in and around Raqqa ranged between 40,000 and 50,000. For further estimates see: Joseph V. Micallef, 'Sitrep Raqqa: The Geopolitics of Eastern Syria', *Military.Com*, 26/6/2017, Accessed on 14/4/2020, at: https://bit.ly/34BMvNK; Tim Lister, 'Battle for Raqqa: Seven Things you Need to Know', *CNN*, 6/6/2017, Accessed on 14/4/2020, at: https://cnn.it/34KV1uh

123. Ibid.

124. The Women's Protection Units (YPJ) is the YPG's female brigade which was set up in 2012. It had a combat role in liberating Raqqa from IS.

125. Joseph V. Micallef, 'Sitrep Raqqa: The Geopolitics of Eastern Syria', *Military.Com*, 26/6/2017, Accessed on 14/4/2020, at: https://bit.ly/2Rzq7PO

126. Ibid.

127. Shawn Snow, 'These Marines in Syria Fired More Artillery than any Battalion Since Vietnam', *Marine Corps Time*, 6/2/2018, Accessed on 14/4/2020, at: https://bit.ly/2ybszW0; Todd South, '3rd Cavalry Regiment Soldiers are Firing Intense Artillery Missions into Syria with Iraqi, French Allies' *Army Times*, 11/12/ 2018, Accessed on 14/4/2020, at: https://bit.ly/3cmZyWl

128. Alex Hopkins, 'International Airstrikes and Civilian Casualty Claims in Iraq and Syria', *Airwars Reports* (July 2017), Accessed on 14/4/2020, at: https://bit.ly/2wCq7HE

129. For the details see: *Wilayat al-Raqqa. 'Inghimassiyun* and Martyrdom Operations', [in Arabic] *al-Naba'*, issue n. 86, Hijri-dated 27 Ramadan 1438, p. 6.

130. Ibid.; Jalal S. Former SDF fighter. Skype interview by author, December 2018.

131. 'ISIS is Digging Underground Headquarters to Avoid Airstrikes in Raqqa', *Syriadirect*, 5/10/2015, Accessed on 14/4/2018, at: https://syriadirect.org/ar/news/مظنت-ةلودلا-رفحت-تارقم-تحت-ضرألا-ج-تيل/

132. Jalal S., Former SDF fighter, Skype interview by author, December 2018; See also: 'After the Battle of Greater Raqqa', *Syrian Observatory for Human Rights*, 18/6/2017, Accessed on 14/4/2020, at: https://bit.ly/3b8V6Kw

133. Jalal S., Former SDF fighter, Skype interview by author, December 2018; Saad K, Former SDF fighter, Skype interview by author, January 2019.

134. It is unclear to the author precisely when the command and control structures of IS Raqqa Province collapsed. However, it was clear that these structures had collapsed by early September 2017. By then, IS units in Raqqa were fighting back aggressively, but without any meaningful coordination. Some of these units started negotiating a safe passage to their elements. Others kept on fighting till the very end of the battle or surrendered without negotiations.

135. 'Al-Raqqa City is ISIS-Free', *Syrian Observatory for Human Rights*, 21/9/2017, Accessed on 14/4/2020, at: https://bit.ly/2VvSXSF; 'Islamic State Faces Endgame in Raqqa, says SDF', *Middle East Eye*, 20/9/2017, Accessed on 14/4/2020, at: https://bit.ly/3eh3s4R

136. Wladimir van Wilgenburg, 'Raqqa in Ruins: Brutal Fight against IS Leaves City Destroyed', *Middle East Eye*, 28/9/2017, Accessed on 14/4/2020, at: https://bit.ly/3ci67ZY; Wladimir Van Wilgenburg, 'Raqqa: IS Sows Chaos as Suicide and Tunnel Attacks Blur Frontlines', *Middle East Eye*, 4/8/2017, Accessed on 14/4/2020, at: https://bit.ly/3ektFzu

137. Wilayat al-Raqqa, '120 Fallen in a Wide-Scale Attack', [in Arabic] *al-Naba'*, issue n. 99, Hijri-dated 8 Muharram 1439, p. 14.

138. *Na'im* (bliss/heaven) and *jahim* (inferno/hell) rhymes in Arabic. The bitterly ironic contrast did not escape the SDF fighters.

139. Stephen J. Townsend, 'Remarks by General Townsend in a Media Availability in Baghdad, Iraq', *US Department of Defense*, 11/7/2017, Accessed on 14/4/2020, at: https://bit.ly/2K4Vtd2

140. Wilayat al-Raqqa, 'We Either Eradicate the Polytheists or Die Trying: There is no Third Choice', [in Arabic] *Al-Naba'*, issue n. 84, Hijri-dated 13 Ramadan 1438, p. 3.

141. There are many conspiracy theories about that incident, including that the US Forces collaborated with IS, and that the SDF manipulated IS units to their benefit. None of these conspiracy theories are credible. The SDF is a coalition of militias that were war-weary and seriously bloodied after about a year of fighting in Raqqa Governorate and four months of fighting in Raqqa City. The compromise came as a way to avoid mutiny and further defections among its own ranks.

142. Sommerville and Riam Dalati, 'Raqqa's Dirty Secret'.

143. Ibid.

144. Evans and Coskun, 'Defector Says Thousands of Islamic State Fighters Left Raqqa in Secret Deal'; Tom O'Connor, 'US Made Secret Deal with ISIS to Let Thousands of Fighters Flee Raqqa to Battle Assad in Syria, Former Ally Says', *Newsweek*, 12/8/17, Accessed on 14/4/2020, at: https://bit.ly/2yTWLoU

The interviewed YPG defector made a series of claims that were all proven inaccurate. Other propaganda claims were mere with the intention of smearing the SDF and the YPG, and dismissing their efforts in fighting IS. Hence, the '3,000' figure is likely to be exaggerated. The figure of 342 fighters and thousands of IS family members

was given to the author by other observers (who were not involved in combat).
145. This happened after capturing the roundabout, the national hospital and the provincial stadium; the last three IS holdouts.
146. 'Kurdish Fighters Raise Flag of PKK Leader in Centre of Raqqa', *Middle East Eye*, 20/10/2017, Accessed on 14/4/42020, at: https://bit.ly/2XxmSw7
147. Pronounced as 'skulls'. See the previous chapter on the Iraqi battlefronts.
148. IS used other tactics including armoured fighting vehicles. But they were relatively limited and not as often relied or reported on by IS media outlets.
149. This dataset is a part of the 'Islamic State Ways of Warfare' Database (ISWD-Raqqa17). The categories of tactics were gleaned from *al-Naba'* newsletter issues numbered from 54 to 103 and IS video releases and photographic reports covering the battlefronts of Raqqa. These are only the reported usage, demonstrating a sample of the quality and quantity of IS tactics. However, many others were gone unreported (see Introduction for the details of the research design).
150. For an excellent review see: Hugo Kamaan, 'Islamic State SVBIED Development since 2014', paper presented at the annual conference of the Strategic Studies Unit entitled 'Militias and Armies: Developments of Combat Capacities of Armed Non-State and State Actors', Arab Centre for Research and Policy Studies (ACRPS), Doha, 24/02/2020, accessed on 9/04/2019 at: https://bit.ly/34BLZzG
151. Saad al-Shari', Syrian activist, interview by author, Gaziantep, 11 November 2018.
152. Ibid. Abu Rajab, Former FSA Fighter who fought ISIS/IS in Idlib, Aleppo and Raqqa, Interview by author, Istanbul, 9 October 2016.
153. Ibid. They were also reportedly used in Deir Ezzor and Aleppo and in Ramadi and Mosul in Iraq (general in Nineveh and Anbar Governorates).
154. For details, see Chapter 1, under the section 'The Ramadi Battlefront: May 2015–February 2016'.
155. Islamic State Ways of Warfare Database (ISWD-Raqqa17). As with other categories, many more may have gone unreported by IS, the media, anti-IS ASAs or ANSAs.
156. As mentioned earlier, the documentation of operations and categories of tactics of IS Provinces in Iraq and Egypt was much more consistent than IS Provinces in Syria and Libya. An example would be the comparative SVBIED figures in Mosul and Raqqa. IS released at least 160 videos or pictures showing SVBIED attacks in Mosul, compared to fewer than 30 for Raqqa.
157. Hamzeh al-Mustapha, 'Combat Performance of al-Nusra Front in the Syrian Civil War', [in Arabic] paper presented at the annual conference of the Strategic Studies Unit entitled 'Militias and Armies: Developments of Combat Capacities of Armed Non-State and State Actors', Arab Centre for Research and Policy Studies (ACRPS), Doha, 23/02/2020, Accessed on 14/4/2020, at: https://bit.ly/2RBCTO5

158. A. W. Former FSA Commander in Raqqa, Interview by author, Istanbul, 9/10/2016. See also: 'Fear of Islamic State suicide attacks lingers in Raqqa', *Sky News*, 22/10/2017, Accessed on 14/4/2018, at: https://bit.ly/2K4Hvbm; Alessandro Rota, 'From Teddy Bears To Bombs: the IEDs of Isis – in Pictures', *The Guardian*, 29/10/2016, Accessed on 14/4/2020, at: https://bit.ly/3cerPy4

159. This is a formal military engineering academy in Aleppo.

160. Interview by author, Gaziantep, November 2018.

161. Ryan Dillon and Eric J. Pahon, 'Department of Defense Press Briefing by Colonel Dillon via Teleconference from Baghdad, Iraq', *US Department of Defense*, 3/8/2017, Accessed on 14/4/2020, at: https://bit.ly/2V5uVyM

162. Alex Horton, 'ISIS Fighters Booby-Trapped Corpses, Toys and a Teddy Bear in Besieged Raqqa', *The Washington Post*, 18/10/2017, Accessed on 14/4/2020, at: https://wapo.st/3c95BO2

163. The 'hedgehog' defensive strong points were placed in the city centre of Raqqa, including *al-Na'im* roundabout. It broke the momentum of the SDF advancement and raised the costs of capturing the area to an unacceptable level for the SDF; hence the negotiations ensued.

164. 'Swarming' in this context is defined as 'engaging an adversary from all directions simultaneously, either with fire or in force'. See John Arquilla and David Ronfeldt, *Swarming and the Future of Conflict*, Santa Monica: RAND Corporation, 2000, p. vii.

165. For the Hasaka campaigns, see for example: Barfi, 'The Military Doctrine of the Islamic State and the Limits of Ba'athist Influence'.

166. There were exceptions due to the concealment tactics and also due to using motorcycles and S/MBIEDs for IS units' mobility in Raqqa.

167. Jalal S., Former SDF fighter, Skype interview by author, December 2018; Saad K., Former SDF fighter, Skype interview by author, January 2019.

168. Ibid.

169. See Chapter 1 for more details on the Iraqi battlefronts and Chapter 3 on the Libyan battlefronts.

170. Ryan Dillon and Eric J. Pahon, 'Department of Defense Press Briefing by Colonel Dillon via Teleconference from Baghdad, Iraq'.

171. Saad K., Former SDF fighter, Skype interview by author, January 2019.

172. Firing in front or behind the target and then dividing the distance until the shelling is directly on target.

173. The relatively slower technique of firing closer to the target with each round.

174. IS was probably unable to use this type of guidance system in the 2017 Battle of Raqqa, due to the heavy presence of jamming devices. However, this simple-but-effective system was used by IS units in 2014 and 2015 in Raqqa and in Deir Ezzor, and by ISIS earlier in 2013 in Aleppo. A. W. Former FSA Commander in Raqqa, Interview by author, Istanbul, 9 October 2016; Abu Qutaybah. Former FSA Commander, Interview by author, Gaziantep, 12 November 2018.

175. Mayssa Awad and James Andre, 'Exclusive: IS group's Armoured Drones Attack from the Skies in Battle for Raqqa', *France 24*, 26/6/2017, Accessed on 14/4/2020, at: https://bit.ly/3ba9aDk
176. Ibid.
177. Martin Smith, 'The Rise of ISIS', *Frontline*, 28/10/2014, Accessed on 14/4/2020, at: https://to.pbs.org/2RzfIUm
178. Saad S., Former FSA fighter in Raqqa and Deir Ezzor, Interview by author, Gaziantep, 12/11/2018; Abu al-Harith. Former Army of Islam – North Commander, Interview by author, Gaziantep, 11/11/2018.
179. Ibid; see also: Reuter, 'Secret Files Reveal the Structure of Islamic State'.
180. Ibid; A. W. Former FSA Commander in Raqqa, Interview by author, Istanbul, 9 October 2016; see Loulla-Mae Eleftheriou-Smith, 'Haji Bakr: Former Saddam Hussein Spy is Mastermind Behind ISIS Takeover of Northern Syria and Push into Iraq, Report Claims', *Independent*, 20/4/2015, Accessed on 14/4/2020, at: https://bit.ly/2XzMokk
181. 'It will be a Fire that Burns the Cross and its People in Raqqah', *Rumiyah*, issue n. 12, 6 August 2017, p. 31.
182. Ground- and air-force units of state armies that engaged ISIS/IS in Deir Ezzor included: the United States, United Kingdom, France, Denmark, Netherlands, Australia, Russia, Iran, Iraq and Syria (Assad's regime). The armed nonstate actors included: Lebanese Hizbullah, Iraqi Popular Mobilisation Units, SDF, Syrian National Defense Forces, Afghani Fatemiyoun Brigade, Pakistani Zainebiyoun Brigade and at least seven other Arab and international organisations.
183. Ellen Francis and Issam Abdallah, 'Islamic State Deploys Car Bombs in Defense of Last Enclave', *Reuters*, 3/3/2019, Accessed on 14/4/2020, at: https://reut.rs/2yfMvHa; Hugo Kaaman, 'From Hajin to Baghouz – Islamic State SVBIED Design & Use', *Hugo Kaaman Open Resource Research on SVBIEDs*, 3/8/2019, Accessed on 14/3/2020, at: https://bit.ly/3eiW0Gc
184. Some of these strikes were in Iraq as well, for details see: Shawn Snow, 'These Marines in Syria Fired More Artillery Than any Battalion Since Vietnam', *Marine Corps Time*, 6/2/2018, Accessed on 14/4/2020, at: https://bit.ly/2V848BL
185. Ibid; also see: Todd South, '3rd Cavalry Regiment Soldiers are Firing Intense Artillery Missions into Syria with Iraqi, French allies', *Army Times*, 11/12/2018, Accessed on 14/4/2020, at: https://bit.ly/34BS9zq
186. Kayla Mueller was a human rights activist and a humanitarian aid worker who provided assistance for Syrian refugees. She was kidnapped in Aleppo in August 2013. She was held, raped and killed by IS.
187. Killed on 27 October 2019 in 'Ayn al-Bayda near Jarablus in Aleppo Governorate. See the full statement in *al-Naba'* newsletter, issue n. 215, Hijri-dated 7 Gumada al-Akhr 1441, p. 4.
188. See: Aymenn Jawad Al-Tamimi, 'The Islamic State's 'Revenge' Expedition for Abu Bakr al-Baghdadi and Abu al-Hassan al-Muhajir: Data and Analysis', *aymennjawad.org*, 31/12/2019, Accessed on 14/4/2020, at: https://bit.ly/2V5w3Cw

189. 'ISIS Appears Again in the Countryside of Raqqa, Attacking the Regime', [in Arabic] *Enab Baladi*, 14/1/2020, Accessed on 14/4/2020, at: https://bit.ly/2wBGHHB; 'Ten Operations by ISIS in Syria Since the Beginning of 2020', [in Arabic] *Enab Baladi*, 15/1/2020, Accessed on 14/4/2020, at: https://bit.ly/3a9H0Hi
190. 'Assassinations' Harvest in the *al-Khayr* [Deir Ezzor] Areas of *al-Sham* [Levant] Province', [in Arabic] *al-Naba'*, issue n. 226, Hijri-dated 24 Rajab 1441, p. 12.
191. One of the most notable examples is the killing of Hajji Bakr during the collective offensive of Syrian armed opposition organisations on ISIS in January 2014.
192. See the following chapter on Libya as an example.
193. For the details of the theoretical argument see Introduction.

4

Reloads, but Implodes: How the 'Islamic State' Fights in Libya

The armoured personnel carriers, the tanks are not their [IS] strongest weapons.

Their strongest weapons are the land mines, booby-traps and the snipers.

Mohamed al-Darrat, Central Axis Commander,
Operation Impenetrable Wall, Sirte.[1]

The beginnings of the 'Islamic State' in Libya

On 22 June 2014, a small local organisation calling itself 'The Islamic Youth *Shura* Council' (IYSC) in the remote eastern town of Derna declared its support for ISIS and its leader. As a result, the first organised ISIS loyalist entity in Libya was established. This development happened only a week before ISIS declared itself as *the* 'Islamic State' (IS) and a 'region-wide Caliphate'. The IYSC was still a new organisation on the Libyan map of armed movements, having publicly declared its existence only on 4 April 2014. Since the early summer of 2014, IYSC intermittently skirmished with other armed groups in Derna including Abu Salim Martyrs Battalion, former Libyan army officers who joined the revolution against Gaddafi's regime, and a faction from *Ansar al-Shari'a* – Derna (Supporters of *Shari'a* – Derna).

While the IYSC failed to take control of Derna, IS still absorbed it and acknowledged it as an official 'province' on 3 October 2014. IS claimed that the areas controlled by the IYSC as *Wilayat Darnah* (Derna Province). On 13 November 2014, Abu Bakr al-Baghdadi declared in an audio statement the establishment of three new 'provinces' in Libya: *Wilayat Barqa* (Cyrenaica Province), *Wilayat Tarablus* (Tripolitania Province) and *Wilayat Fazzan* (Fezzan Province).[2] Derna

Province was the first nucleus of the Cyrenaica Province that encom-passed parts of neighbourhoods in Benghazi and Derna. In early 2015, IS in Libya established some territorial control and governance mecha-nisms in parts of Derna, Benghazi and Sabratha. By 31 May 2015, IS had occupied Sirte. However, by July 2015, IS had already been defeated in Derna's town centre by a coalition of local armed groups and former Libyan army officers. IS in Libya also lost its leader, Abu Nabil al-Anbari, to an American airstrike in November 2015.[3] A year later, in December 2016, IS lost Sirte to forces loyal to the Libyan government in Tripoli, backed by international military support. As of 2017, IS has mainly retreated to the south of Libya and shifted back to hit-and-run/non-fixed-position guerrilla tactics and terrorism.

This chapter provides a brief historical overview of the birth of IS in Libya, as of 22 June 2014. It aims to explain how IS was able to gradually develop their combat capacities in Libya since then. As a result of this, IS was able to take control of parts of Derna in October 2014, and the whole of Sirte by the end of May 2015. The occupa-tions happened despite a lack of local support, state sponsorship or supportive geography.[4] The chapter focuses on analysing the battle-fronts of Derna and Sirte between June 2015[5] and December 2016, as a sample reflecting how IS fights in Libya. The chapter is partly based on interviews with soldiers and militiamen who fought against IS in the aforementioned battlefronts. It is also based on documents produced by IS in Libya, represented by two of its three 'provinces': Cyrenaica and Tripolitania. The chapter also relies on official documents released by the US government, and on other open-source materials.

The chapter is composed of six sections. Following this first sec-tion, the second section overviews the military build-up of IS in Libya since June 2014. The third and fourth sections outline the details of the abovementioned battlefronts of Derna and Sirte within specific timeframes.[6] The fifth section analyses how IS fights in Libya, using data and observations from the two battlefronts and elsewhere in Libya. Finally, the concluding section reflects on the future of IS insurgency in Libya, after losing territory and shifting back to guer-rilla and terrorism ways of warfare.

The IS military build-up in Libya: an overview

The IS military build-up in Libya had a similar trajectory to that of Syria.[7] It was also based on the iALLTR *modus operandi* of intelligence-gathering; absorbing already-existing, like-minded organisations, and upsizing them by individual and collective recruitment of youth;[8] looting parts of the *ancein régime*'s arsenals; relying on the experiences of IS commanders/fighters from combat zones abroad to lead and guide local affiliates; and rapid transfer of knowhows of tactical innovations, strategic shifts and operational knowledge relevant to combat effectiveness.

Initially, a number of Libyan fighters who had fought in both Iraq (2004–12) and Syria (2012–14) returned back to their hometown of Derna. These individuals established the connection between ISIS in Syria and the IYSC.[9] One of the Libyan founders was a TWJ/AQI/ISI member who had fought in Iraq: Hasan al-Saliheen al-Sha'iri.[10] He is also known as 'Abu Habiba' and 'Al-A'war' (the one-eyed) due to the loss of his left eye in a combat injury in Iraq.[11]

In terms of manpower, IS in Derna partly relied on veterans of wars in Iraq and Syria. Some of those veterans were affiliated with various armed formations including 'al-Battar Battalion', a formation that fought against the Assad regime in 2012 before many of its fighters joined ISIS and, later on, IS. More 'knowhow-transfers' and experienced commanders arrived in Libya since September 2014, including figures as senior as Abu Nabil al-Anbari (Wissam al-Zubaydi),[12] who became the IS governor (*wali*) of Libya after holding the position of IS governor of Saladin Province in Iraq. Al-Anbari was killed in an American airstrike on Derna on 13 November 2015.[13] According to his eulogy, he militarily led the IS offensive on Samarra City in Iraq (May–June 2014) and played a leading role during the fall of Baiji and Tikrit in June 2014.[14] In addition to experienced fighters, IS in Derna also recruited young local jihadists with limited-to-no combat experience, as well as defectors from both armed Islamist groups and the regular army of the former dictator, Muammar Gaddafi.[15]

In November 2014, IS in Libya was still a relatively small organisation with an estimated manpower of 400 fighters concentrated in eastern Libya (mainly in Derna and Benghazi), with scattered cells across the large, unpopulated country.[16] IS territorial control efforts changed that. After the fall of Sirte on 31 May 2015, the United Nations estimated IS manpower in Libya to number between 2,000 and 3,000 fighters in December 2015.[17] At a minimum, IS manpower grew five times despite their defeat in Derna in July 2015[18] and the decapitation of their leader, Abu Nabil al-Anbari, in November 2015. In June 2016, six months before losing Sirte to *al-Bunyan al-Marsus*[19]

(BM) forces, the then head of the Central Intelligence Agency, John Brennan, told the US Senate Intelligence Committee that the fighting force of IS in Libya ranged between 5,000 and 8,000 fighters.[20] In 2018, the US Africa Command (AFRICOM) estimated the number of IS fighters in Libya to be around 750 fighters.[21]

In terms of armament and looting existing arsenals, the IYSC revealed around 30 technicals in a military parade in April 2014 after publicly declaring its existence.[22] The weapons shown in the parade were a mix of variants of AK-47 assault rifles, variants of PK all-purpose machine guns and at least one single-barrelled ZPU-1 anti-aircraft autocannon (mounting one KPV, 14.5mm heavy machine gun). The latter was the heaviest weapon displayed in the April 2014 parade. In an October 2014 parade, the number of technicals almost doubled. About 60 technicals were shown during that month, after the IYSC declared its support to IS.

As Ali bin Gharbiyah, the assistant commander of the Eastern Axis in the *al-Bunyan al-Marsus* (Impenetrable Wall) Operation, which fought IS in Sirt, said:

> The organisation's weapons are taken from the remnants of the Libyan army's arsenals . . . they reached *Daesh* [ISIS][23] after [factions of] *Ansar al-Sharia* in Derna, Benghazi and Sirte pledged allegiance to the organisation.[24]

The conventional weaponry taken from the former Libyan army's arsenal did not grant IS any tactical edges over other armed formations in Derna and Sirte, especially with the limited manpower of the organisation. However, 'the car bombs and explosive belts were the distinct weapons that the organisation (IS) had. These were easy to manufacture from the remnants of damaged shells and munitions', said Gharbiyah.[25]

The IED and sniping-intensive tactics were steadily developing in Libya, mainly via 'knowhow-transfers' from other IS provinces in Iraq and Syria. In Derna, between 5 January 2016 and 19 April 2016, IS has claimed that it used IED-intensive tactics 32 times and sniping-intensive tactics 10 times.[26] The breakup of the 32 IED-intensive tactics yielded 2 SVBIEDs, 1 VBIED and 29 static IEDs. The two categories of tactics (IEDs and sniping) were used more than any other category in Derna during the same period. In Sirte, the ISWD database recorded IS usage of IED-intensive tactics 99 times (63 static IEDs and 36 SVBIEDs), and sniping-intensive 16 times between January and December 2016.

IS had also used S/VBIEDs in Libya, both in urban terrorism and as a battlefield weapon in conventional and guerrilla operations. Many of the SVBIEDs used by IS in the Cyrenaica and Tripolitania Provinces were pick-up trucks of similar design. This may reflect the organised 'knowhow-transfers' across eastern and western Libya.

IS in Libya has also shown effectiveness in using guerrilla formations (GFs), and suicidal guerrilla formations (SGFs) or *inghimassiyun* (plural of *inghimasi*) in urban terrorism and conventional operations. The SGFs attacks on the Corinthian Hotel in Tripoli in April 2015,[27] and on the Courts Complex in Misrata in October 2017 are examples of IS urban terrorism operations using SGFs, before and after the defeat in Sirte.[28] In terms of conventional warfare, IS in Libya had relatively limited numbers of armoured vehicles, most of which were destroyed by airstrikes in Sirte or by IEDs and anti-tank mines and rockets in Derna. Compared to IS provinces in Iraq and Syria, IS in Libya had limited usage for armoured and artillery.[29] The limited quantity of armoured vehicles did not inhibit IS qualitative innovations. In March 2016, during the battle of Derna between IS units and local armed groups,[30] IS launched a counteroffensive with a converted infantry fighting vehicle made of a 6x6 truck and a BMP-1 hull with metal plating, slat, and BMP-1 turret. The turret housed what looked like an SPG-9 recoilless gun (73mm). Assault rifles and a PKS machine gun fired out of the BMP's firing ports. Briefly, IS units had a turned a 6x6 truck into an armoured large technical while they were under siege in Derna.[31]

Overall – in less than a year – IS in Libya had built-up military capabilities that transformed it from a small armed organisation that failed to take over the remote town of Derna to a mid-size armed organisation with a capacity for executing urban terrorism and guerrilla operations across eastern and western Libya, in addition to limited-but-innovative capacity in quasi-conventional warfare. These hybrid capacities were utilised to occupy the city of Sirte from March 2015 to December 2016.

The Battlefront of Derna: June 2015–April 2016

As mentioned above, the 'Derna Province' of IS was declared in October 2014. The following month, al-Baghdadi declared the establishment of 'Cyrenaica Province' in an audio statement. In response, the '*Mujahidin* and Revolutionaries *Shura* Council of Derna and its

Environs' (Consultative Council of Holy Fighters and Revolutionaries of Derna and Its Environs or DMSC)[32] was established on 12
December 2014. It comprised almost all anti-IS local armed organisations and factions, in addition to some of the Derna-based Libyan
army officers. The DMSC also formed transregional alliances with
anti-IS armed formations from al-Bayda, Tobruk and other eastern
towns, most notably the Ali Hassan al-Jaber battalion led by Colonel
Muhammad Boughifayir.[33] The dual objective of this alliance was to
fight IS in Derna[34] and to stop Khalifa Hefter's self-declared 'Libyan
National Army'[35] (LNA) from storming Derna and its suburbs.

There was a history of intermittent clashes between the predecessor of IS in Derna (the IYSC) and the DMSC's components, especially the Abu Salim Martyrs Battalion. These clashes were sparked
by three main local issues. The first was the Battalion's protection
of Mustafa Abdul Jalil, the former Chairman of the National Tran-
sitional Council (NTC), considered by the IYSC as a 'leader of an
apostate entity'.[36] The second issue concerned the battalion's sign-
ing of a contract with the Libyan Ministry of Interior and therefore
becoming a part of the (defunct) Supreme Security Committee.[37]
This was considered by the IYSC as joining 'apostate' entities that
do not implement 'the laws of God'.[38] The third issue was the protection of the July 2012 electoral centres in Derna by Abu Salim
Martyrs Battalion, an 'accusation' which the battalion leaders gave
mixed messages about.[39] IYSC considered this as offering protection
for 'apostates' implementing the 'idolatry practices of democracy'.[40]
These ideological-political local disputes between local armed organisations were capitalised upon by ISIS/IS.

The initial skirmishes were sparked by the assassination of the
Abu Salim Martyrs Battalion's ideologue, Mohammad Boubilal,
on 30 May 2014 via booby-trapping his car. By October 2014,
IYSC/IS was unsuccessfully employing a common tactic known in
IS literature as '*tatwiq wa tadyiq*' (encircle and choke). This is a form
of siege warfare, where the organisation besieges a town/city and
advances gradually, only after cutting or mining all supply routes
between its foes' units and to them. The tactic was attempted on
Deir Ezzor City in July 2014 and on the town of Kobane in September 2014. In Derna, IS and their local allies did not have enough
manpower to execute 'encircle and choke'. Still, they attempted it.
They had at least three small camps around the town, all used to station technicals, and drove over thirty of them to block the entrances

of the town and relevant routes within it. By December 2014 – after al-Baghdadi's declaration in his 13 November audio statement – IS formations were well-positioned in the town centre. Between January and June 2015, the skirmishes gradually escalated from a campaign of booby-traps, sniping and close-quarter assassinations to a full-fledged urban battle with frontlines in Derna's centre and its (mainly eastern) suburbs.

Overall, the DMSC and its allies from other Libyan towns launched three main operations against IS in Derna. The first was fought in an urban terrain and mockingly codenamed '*hal laha min baqiya?*' (Does It [IS] Have Any Remnants Left?). The DMSC was mocking the IS motto '*baqiya wa tatamddad*' [remaining/enduring and expanding] by using a verse from the Quran (Chapter 69 - *Al-Haaqqa* [The Reality]: Verse 8) about a group of oppressive sinners destroyed by the wrath of God. The battle was fought between 9 June 2014 and 1 July 2014 in the town centre. The second operation was fought in a mix of urban and suburban terrains. It was codenamed '*al-Nahrawan*'[41] by the DMSC. The DMSC and its allies targeted the retreating IS units in coastal east Derna and the eastern suburbs, such as al-Korfat, Lehjjaj and al-Fattaih areas. By 10 July 2015, the DMSC and its allies were victoriously advancing, except in al-Fattaih. The third and final operation was codenamed 'Martyrs of al-Qubba'.[42] Its battles raged in al-Fattaih area and al-Hila juncture from 15 November 2015 until 20 April 2016. On that date, the DMSC and its allies declared that Derna was liberated from IS. On Derna's battlefields, the DMSC had allegedly killed over 300 IS fighters and captured tens of them.

In June 2015, the manpower of IS in Derna was estimated to range between 500 and 800 fighters at the very maximum. They were mainly concentrated in the centre and the east of the town, from the coastal Sahel area in the north to the south of al-Fattaih area. IS had significant presence around al-Hareesh Hospital (controlled by IS) all the way to the Pearl Hotel (temporal command centre of IS in Derna),[43] in addition to *al-Arba'mi'a* (The 400) neighbourhood and the strategic al-Hila area (containing a road juncture with access to the south and to the east of Libya).[44]

In terms of armament, IS in Derna was relying on a common guerrilla and terrorism arsenal: handguns (with and without silencers), assault rifles, sniping rifles, medium and heavy machine-guns, light and medium mortars, IEDs and technicals. Additionally, a few conventional capacities existed among IS units in Derna, including at least seven locally modified armoured fighting vehicles and at least one T-55 tank.[45]

On 9 and 10 June 2015, two DMSC leaders, Naser al-ʿiker (deputy head of the Consultative Council) and Salim Derby (the Commander of Abu Salim Battalion and the Head of the Council) were killed in confrontations with IS. The first was assassinated with his assistant, Faraj al-Houti, by IS gunmen in al-Shiha area of Derna on 9 June. In retaliation, Derby led an attack on an IS position in western Derna, where IS forces were limited. He was killed during the attack on 10 June. Both figures were relatively popular in Derna. Derby, specifically, had been calling for a general de-escalation and national reconciliation among all Libyans, including Gaddafi's loyalists and others.[46] His death, among others, coupled with IS aggressive behaviour, led to a public backlash, including rallies and demonstrations, against the IS presence in Derna.[47]

While decapitated, the DMSC and their allies managed to escalate the campaign by targeting nine IS positions in Derna almost simultaneously in the following week. Under sustained military pressure, IS reacted with a wave of SVBIEDs and SVs. 'Their operations were unconventional. It involved all types of boobytrapping and mining . . . it certainly delayed their defeat,' said Colonel Ismail Shukri, a commander in the Libyan Military Intelligence branch in Misrata who was closely monitoring the Derna battlefront.[48] By 1 July 2015, DMSC was decisively victorious in the centre of the town. IS still controlled parts of the east coastal suburbs, all of al-Fattaih in the southeast of Derna, and the critical al-Hila road juncture from which the organisation accessed logistical resources from loyalists in Ajdabiya and Sirte.

The anti-IS local coalition of armed groups began to heavily rely on guerrilla tactics. According to Colonel Muftah Hamza, the Commander of al-Hila frontline against IS in Derna:

> We had to terrorise them like they terrorised us. We would hit-and-run at night behind their lines. We planted mines to sabotage their armoured vehicles and planted IEDs on their routes . . . all the youth [armed young men] in Derna were helping us.[49]

On 11 July 2015, IS issued a propaganda video entitled 'Patience, Derna's Awakenings'.[50] The organisation implicitly acknowledged its military defeat in Derna, sent support messages to IS prisoners held by the DMSC, and promised the latter more IEDs.[51] Despite the threats, IS remained beleaguered in al-Fattaih and could not expand.

On 13 November 2015, IS in Libya lost its leader – Abu Nabil al-Anbari – in an American airstrike.[52] Following his death, IS in Derna escalated its campaign. Between November 2015 and February 2016, it took responsibility for 15 sniping-intensive attacks, 32 IED-intensive attacks and seven SGF operations.[53] IS in Derna also succeeded in killing two leading figures in the coalition against it: Colonel Mohammad Boughifayir, the commander of al-Fattaih frontline until his death on 15 November 2015, and Colonel Salih Sahd, his deputy and one of Derna's leading army officers. Both were east-based military officers who had defected from Gaddafi's army to join the February revolution and, at a later stage, refused to join Khalifa Hefter's 'Libyan National Army' (LNA) coalition of tribal militias and army defectors.[54] Both colonels were killed on the al-Fattaih frontline while coordinating an anti-IS operation with the DMSC and other armed units. 'We lost more than 215 of our men . . . mostly in [IS] sniping attacks and night raids', said Colonel Muftah Hamza, who commanded the units in al-Fattaih after Boughifayir and Sahad.[55] 'But we hurt them . . . they had about 100 fighters left in al-Fattaih. They were led by a Libyan – Ibrahim al-Warfalli – but I think most of the fighters were foreigners', said Colonel Hamza.[56]

The Iraqi Abu Nabil Al-Anbari was succeeded by the Saudi Abdul Qadir al-Najdi, who threatened to 'invade Rome' during an infamous propaganda interview with *al-Naba'* newsletter published on 8 March 2016.[57] Rather than 'invading Rome', al-Najdi decided to fully retreat from Derna's al-Fattaih area to Sirte on the late night of 19 April 2016 and the early hours of 20 April 2016. 'They escaped at night . . . mainly through areas and routes controlled by the Dignity Operation [LNA militias] and without any interception from them', said Colonel Hamza.[58] The IS force that retreated from Derna was estimated to be 70 armed vehicles (mainly soft-skinned pick-trucks and technicals) carrying between 217 and 300 individuals (a mix of IS fighters and a few families).[59] Their destination was Sirte. 'We knew them by name after [the 2016 battle of] Sirte. They were 217 fighters. We just wonder why they weren't bombed by Hefter's forces in an 800-kilometre [500 miles] journey from Derna to Sirte', said Brigadier-General Muhammad Al-Ghusri, a commander in the Libyan army units of the internationally recognised Government of National Accord (GNA) and the spokesperson of *al-Bunyan al-Marsus* Operation to liberate Sirte from IS.[60]

The Battlefront of Sirte: March 2015–December 2016

While IS was gradually retreating in the battles of Derna, the organisation was gradually advancing in the more strategic battles for Sirte.[61] Similar to Derna in *Wilayat Barqa* (Cyrenaica Province), Sirte was gradually becoming the centre of *Wilayat Tarablus* (IS' Tripolitania Province). In IS documents, Tripolitania Province extends 465 miles along the Libyan coast, from al-Nufaliya town (90 miles east of Sirte) to Ra's Ajdir town on the Libyan–Tunisian border. However, IS militarily controlled only small parts of this coastal stretch at different timeframes, including al-Nufaliya, Harawa, Abu Grain, Sirte and parts of Sabratha (50 miles west of Tripoli). IS had a presence and used urban terrorism and guerrilla warfare tactics elsewhere in western Libya, including in Tripoli and Misrata.

In Sirte, IS units were already well-positioned in and around the city just before the armed escalation in Derna had started in June 2015. This well-positioning was due to several reasons. On a macro-level, although the socio-political environment in Sirte had a few similarities to that of Derna, there were some differences that helped IS' expansion in the former dictator's birthplace[62] and tribal stronghold.[63] In Sirte, IS was attracting individuals and factions from the local branch of *Ansar al-Shari'a*, the local salafi-jihadist organisation that was already operating in Sirte. The *Ansar* in Sirte were well-established by June 2013. In the following six months, the group had the capacity to organise public events, provide security patrols in various neighbourhoods and in the University of Sirte, undertake some policing work and regulate traffic, arbitrate on issues between tribes and clans in and around Sirte, clean roads and undertake other local governance tasks. But unlike Derna,[64] the pool of potential recruits was wider than defectors from, and the collective absorption of, local jihadist entities. This was mainly due to the particularities and the history of Sirte under Gaddafi's regime. IS did not recruit from the salafi-jihadist pool only, but also from Sirte-based military and civilian Gaddafi loyalists, in a similar fashion to how IS/ISIS/ISI/AQI/TWJ had behaved towards a few Baath Party loyalists in Iraq.[65] In a way, what IS was proposing intersected with many elements of Gaddafi's propaganda and positions. IS was anti-Western, anti-democratic, extremely brutal, highly tolerant of repression and human rights violations and hostile towards the 17th February revolution and its objectives; and so was the Gaddafi regime.[66] 'Some of the former military officers in Sirte preferred *Daesh* [ISIS] to the 17th February revolution. That helped the organisation to advance', said General Youssef al-Mangush, the former Chief of Staff of the Libyan Army.[67] Ahmed Gaddaf al-Dam, Gaddafi's cousin and one of the main leaders and former officials of his regime, publicly declared his support of ISIS on an Egyptian TV channel on 17 January 2015.[68]

During that same month, IS units were already taking positions in Sirte with little-to-no resistance. Dozens of members from local tribes, including from the Qadhadhfa[69] (pronounced Gaddadafa), Warfalla, al-Firjan and Tarhuna tribes assisted in taking control of critical infrastructure.[70] That included using the Ouagadougou Conference Centre as a command-centre for the organisation.[71] From there, the organisation organised an armed occupation of the town of al-Nufaliya with over 40 armed 4x4 vehicles and technicals – all departing from Sirte on 8 February 2015. The al-Nufaliya battle was necessary to act as an eastern strategic buffer for Sirte. And in case Misratan forces attacked from the west, al-Nufaliya (and later Harawa) could act as a strategic depth for an IS withdrawal.

The battle of al-Nufaliya was a quick victory for IS. It was among the few battles in which IS both outgunned and outnumbered its enemies. There were about 50 local policemen left in al-Nufaliya when IS attacked, led by Lt. Col. Mohamed Obaid.[72] IS manpower was estimated to be about 200 fighters boarding 40 4x4 vehicles and technicals, proving superior with an estimated manpower ratio of 4-to-1. After occupying the town, the organisation declared Ali al-Qarqa'i (Abu Hamam al-Libi) as the emir of al-Nufaliya.

Confident after its quick victory in al-Nufaliya, IS units took over the local radio station of Sirte on 12 February 2015. Largely unopposed, IS started broadcasting the speeches of its leader, Abu Bakr Al-Baghdadi, on 13 February 2015 on the Sirte Radio waves.[73] On 19 February 2015, IS was *de facto* unchallenged in Sirte; in a show of force, the organisation organised a military parade of about 60 armed 4x4 vehicles and technicals.[74] IS expanded further in the city to control al-Wataniya television studio, Ibn Sinah Hospital, the University of Sirte, the immigration centre and local government buildings.[75] The organisation installed a local emir in charge, Walid al-Firjani, a relative of a former senior military officer in Gaddafi's regime.[76]

By early March 2015, IS had deployed eight armed formations, each ranging between 50 and 100 fighters, to control all of Sirte.[77] With these developments, Libya Dawn – a coalition of military formations operating under the Libyan army's Chief of Staff, who was still loyal to the National Salvation Government in Tripoli – reacted.[78] Libya Dawn deployed the 166th Battalion around Sirte, in an attempt to liberate the city centre, the university and other areas from IS units. 'IS forces had about 100 to 150 armed vehicles [technicals and 4x4 vehicles]' stated Mohammed Ali Abdullah, an elected member of the General National Congress[79] from Misrata and the leader of the National Front Party.[80] The 166th Battalion and its local allies had around 300 armed 4x4 vehicles and technical carrying 1,200 to 1,500 soldiers and fighters.[81] The battle ensued on 14 March 2015. IS was slightly outnumbered with a manpower ratio of about 1.8-to-1.

The battle was sporadic, chaotic and brutal. The 166th Battalion's ground forces initially attacked in the western suburbs of the city and used aerial bombardment to attack IS command-and-control in the Ouagadougou Conference Centre.[82] By 17 March 2015, the Battalion seemed to be advancing on all fronts in and around Sirte. It announced the killing of 24 IS fighters, the capture of tens and wounding of more than 39, including 'Ajjaj al-Saqr, a high-ranking security officer in Gaddafi's regime who was fighting on the side of IS.[83] Moreover, during its operations to the east of Sirte, Libya Dawn killed the IS emir of al-Nufaliya, Ali al-Qarqa'i, during the battle of Harawa (45 miles to the east of Sirte). The battalion's quick advance, however, turned into a bloody stalemate followed by a full retreat. Both the 166th Battalion and IS in Libya had limited resources in terms of manpower, firepower, external support, funds and ammunition. IS, however, had shown itself to be better at managing limited resources, economising the use of ammunition, disaggregating-reaggregating its units and absorbing initial losses, including the loss of commanders.

Three months into the fight – in early June 2015 – the battle outcome was not looking good. The 166th Battalion had lost both momentum and resources.[84] 'We can stay on the front line eating only bread and drinking only water, but the forces can't stay there without ammunition and weapons', said Mohamed Zadma, a commander in the 166th Battalion in June 2015.[85] With limited resources and fatigue wearing down anti-IS units, IS escalated their activity with a wave of SVBIEDs including an attack on a military checkpoint near Misrata. By early June 2015, IS was advancing in the south, east and west of Sirte and its suburbs.[86] Soliman Mousa, an officer within the 166th Battalion confirmed that his forces had retreated in the face of IS in all four directions;[87] including via the Mediterranean sea to the north by makeshift boats.[88] At this point, IS fighters had captured almost all of Sirte, parts of its suburbs and nearby areas. The fallen areas included strategic locations such as al-Bukhariya station and al-Khamsin in the west, the Qardabiya airbase (12 miles to the south of Sirte), and the barracks of the 136th Battalion (where IS seized a number of T-55 tanks) and the main water treatment facility (part of the Man-Made River Project) – both to the east of Sirte. By mid-June, IS was decisively victorious in the 2015 battle of Sirte.

The war for Sirte was far from over, however. IS advances west of Sirte and its occupation of the strategic town of Abu Grain,[89] alarmed both Tripoli (where the internationally recognised Government of National Accord – GNA – had finally arrived on 30 March 2016) and Misrata (from where the most organised military units loyal to the GNA hail). On 5 May 2016, a GNA-sanctioned military operation was established under the codename *al-Bunyan al-Marsus*[90] (BM); the objective was to liberate Sirte and other western towns and villages. The GNA and the Misratan commanders managed to mobilise around 6,000 soldiers and fighters, with artillery and armour pieces, and the GNA-loyal Libyan air-force units. Critical to the battle outcome, the GNA and the BM forces managed to secure support from the international coalition against IS, mainly from the United States, the United Kingdom and Italy. Three months into the battle, the United States' AFRICOM launched 'Operation Odyssey Lightning' on 1 August 2016 to 'assist the GNA and other coalition partners to remove ISIS fighters from Sirte'.[91] The operation involved the US airforce, Marine Corps Harriers,[92] as well as British Special Air Service (SAS) and Special Boat Service (SBS) units.[93]

On the other end, the manpower of IS Tripolitania Province in Sirte was estimated be at maximum 2,000 fighters. Hence, the organisation was outnumbered 3-to-1. In terms of armament, IS in Sirte had a combined conventional, guerrilla and terrorism arsenal, which included at least four T-72 tanks and five T-55 tanks,[94] at least 12 well-armed guerrilla formations (squad- to company-sized),[95] a number of SA-16 and SA-18 MANPADs and about a thousand IEDs.[96] The amount of IEDs available allowed the organisation to allegedly execute 120 suicide operations between mid-May 2016 and the beginning of December 2016, according to the Emir of Tripolitania Province Abu Hudhayfa al-Muhajir in an interview with *Rumiyah* magazine, one of the IS outlets.[97] Out of the alleged 120 suicide operations, at least 36 of them were SVBIEDs verified and recorded in the ISWD dataset by the author.[98] This is in addition to the SGF operations, which involved at least five suicide vests.[99]

From early to mid-May, the BM forces advanced slowly but consistently from the eastern borders of Misrata District crossing the western borders of Sirte District.[100] The BM forces captured the strategic town of Abu Grain (90 miles west of Sirte) on 17 May, after 12 days of fighting. On 19 May, IS responded with a wave of SVBIEDs and SLVBIEDs slowing down the BM advances, in a common IS momentum-breaking tactic. On that day, 30 soldiers were killed in one SLVBIED operation while they were still 56 miles away from Sirte.[101] The BM forces stormed the western outskirts of Sirte for the first time on 8 June 2016, despite sustaining heavy causalities before and during that operation. The Petroleum Facilities Guards,[102] a GNA-loyal militia supported by the BM, were also advancing from the east.[103] The militia reached Harawa (45 miles east of Sirte) by 8 June 2016.[104]

While the BM and its partners were advancing, IS was still infiltrating and striking behind their lines, including a devastating SVBIED-led attack on the Abu Grain police station that killed 10 soldiers on 16 June 2016.[105] The BM had lost over 200 soldiers in the first month of a six-months battle; or about three per cent of their overall manpower. But the heaviest casualties sustained by the BM were in July 2016, just before the US intervention. The causalities were mainly a result of SVBIEDs and sniper fire. During that month, IS manpower was concentrated in the port area in the north of the city and Ouagadougou Conference Centre in the southeast of it. On 15 July 2016, the BM failed to take over the Ouagadougou Conference Centre after three SVBIEDs hit their attacking force, followed by a counterattack of four squad-sized IS guerrilla formations.[106] The IS counterattack left over 20 BM soldiers dead and over 120 wounded.[107] The BM offensive aiming to take over the port area had also failed in July. The failure was due to a mix of IED and sniping-intensive tactics employed by at least ten squad-sized IS guerrilla formations in the port areas, including in the first and the second neighbourhoods and al-Jiza al-Bahariya areas.[108] During the battles in and around *al-Shatt* (Beach) Road, the BM forces suffered over 40 deaths, and 140 BM fighters were wounded.[109]

Two months into the battle, it was clear that the BM forces could not tolerate the rate of their causalities. 'They [BM units] will never be able to hold that position now', said Brigadier-General Muhammad Al-Ghusri, a spokesman for the GNA/BM forces operating in Sirte.[110] He was referring to the position in the 700 neighbourhood, about three miles away from Ouagadougou Conference Centre. Despite the failures, the BM strikes had downgraded IS in Sirte. By mid-July, the organisation's command-and-control in Sirte had collapsed. 'We learned more about them [in July]. They became uncoordinated and decentralised separate groups fighting [independently] in different parts of the city', said Colonel Shukri.[111]

The US intervention of 1 August 2016 (Operation Odyssey Lightning) came at a critical moment for the BM operation; slowly and significantly further downgrading IS in Sirte. This was achieved through a combination of precision airstrikes, intelligence support and military advisors to the GNA and BM forces, and (American, British and Italian) special forces attacks.[112] On 10 August 2016, ten days after the US intervention, the BM forces managed to finally takeover the Ouagadougou Conference Centre and nearby areas south of the city, including Ibn Sina hospital and the University of Sirte campus.[113] On 12 August 2016, the BM forces, with American assistance, launched another operation in the north and the northeast of Sirte. The operation was codenamed 'Macmadas', an ancient name of Sirte.[114] It was another attempt to liberate the north, including (from west to east) the first and the second neighbourhoods, al-Jiza al-Bahariya and the third neighbourhood. After three months of intense fighting, the BM managed to liberate all of them, including (lastly) al-Jiza al-Bahariya on 6 December 2016.

Overall, AFRICOM announced that it conducted 495 precision airstrikes in Sirte between 1 August and 19 December 2016.[115] AFRICOM estimated that 800 to 900 IS fighters were killed during the operation in Sirte. It was a pyrrhic victory for the BM, the GNA and the American-led international coalition. The Misratan-led forces lost over ten per cent of their soldiers (over 700 deaths), and an estimated one in every two BM fighters in Sirte was wounded during the six-month series of battles.[116]

How IS fights in Libya

Compared to Iraq (Mosul) and Egypt (Sheikh Zuweid) where IS was ludicrously outnumbered, the organisation did not manage to pull off any surprising military upsets on the Libyan battlefields. IS was only slightly outnumbered – though significantly outgunned – in most of the battles it fought in Derna, Sirte and elsewhere in eastern and western Libya such as in Sabratha.[117] Still, IS managed to occupy and then expand in parts of Derna and in all of Sirte between 2014 and 2016. Despite the relatively limited upsets pulled off by Cyrenaica and Tripolitania Provinces, IS in Libya managed to surprise its foes with tactical innovations and strategic shifts (from terrorism to guerrilla to conventional and back) similar to those implemented in Iraq and Syria, albeit of lesser quality. The following section analyses some of the tactical and operational patterns employed by IS in Libya.

Derna: when tactics fail strategy

Parts of Derna were occupied via a slightly modified *modus operandi* compared to how IS took over cities and towns in Iraq and Egypt. In contrast to Fallujah, IS in Derna did not coalition-build and fight, before it liquidated its coalition partners. And unlike Sheikh Zuweid and al-Nufaliya, there was no successful *blitzkrieg*-like[118] decisive charge with technicals in the takeover of Derna. There were some similarities, however, with the ISIS occupation of Raqqa in Syria.[119]

The partially successful attempt on Derna in 2014–15 was a result of operational-level combinations of '*tatwiq wa tadyiq*' (encircle and choke) and 'soften and creep'[120] between November 2014 and June 2015. IS used a combination of IED and sniping-intensive tactics, as well as multiple tactics of close-quarter assassinations. The assassinations, however, undermined the strategic of goal of controlling Derna, as well as the operational goal of beleaguering anti-IS armed groups in different neighbourhoods so that they fail to support each other. Rather than being terrorised into submission or falling into disarray, Derna-based armed groups reacted to the assassinations by collectively attacking all IS positions in the town. Moreover, the DMSC formed tactical alliances with nearby battalions from outside of Derna. IS had to fight defensively, without prior offensive successes to control the town and raid its resources. Table 4.1 (overleaf) outlines a sample of reported tactics employed by IS in Derna and recorded in the ISWD dataset between January and April 2016.[121]

Table 4.1 IS Reported Tactics in Derna, January to April 2016[122]

Category of Tactics	Reported Usage
IED-intensive	29
Sniping-intensive	10
Close-quarter Assassinations	5
Suicidal Guerrilla Formations (SGFs)	5
Artillery	4
Unguided Rockets	3
Guerrilla Formations (GFs)	3
SVBIED-intensive	2
ATGMs	2
VBIED-intensive	1
Anti-aircraft Autocannons	1
Total	**65**

IS defensively fought back three urban/suburban offensives for eleven months in Derna. It relied overwhelmingly on the above-mentioned eleven categories of tactics to fight other ANSAs in the town. In January 2016, when al-Naba' newsletter undertook more regular reporting on Derna, the organisation was already losing military momentum. Holed up in al-Fattaih and only four months away from a full retreat, IS in Derna relied mainly on IEDs and sniping tactics. The organisation's units in Derna had limited choices; either to retreat or to be destroyed. The choice was rational: retreat and regroup in Sirte to fight another day.

Sirte: strategic shifts

Operationally, the attempt on Sirte was different from, and more successful compared to, the attempt on Derna. Neither 'encircle and choke' nor 'soften and creep' were the dominant *modus operandi* in Sirte. IS did not build an operational coalition to occupy the city as it did elsewhere. Hence, the operational plan shifted in Sirte.[123] Unlike Derna, the organisation successfully absorbed factions of local armed groups, fractions of local tribes and tens of former Gaddafi officers,[124] rather than starting with a terror campaign against them. This partly succeeded due to the aforementioned environmental conditions and particularities of Sirte. Consequently, IS in Sirte had more resources compared to IS in Derna,[125] including in terms of manpower, fire-power, quality of weapons and quantity of ammunitions. In Sirte, IS used MANPADs,[126] Hbieds and multiple T-55 and T-72 tanks. In Derna, these resources were either unavailable or available in very limited quantities compared to Sirte (such as the case of the T-55 tank in Derna).

In terms of tactics used in Sirte, there was more continuity rather than change – unlike at the operational level. The usual IS reliance on combinations of IED-intensive and sniping-intensive tactics was executed in a similar fashion to Derna. However, IS units in Sirte also used tunnel warfare, tanks and at least 36 SVBIEDs. There was a notable difference in the effectiveness of different categories of tactics though, as observed by BM commanders who fought IS in Sirte. 'The armoured personnel carriers, the tanks are not their [IS] strongest weapons. Their strongest weapons are the land mines, booby-traps and the snipers', observed Mohamed al-Darrat, the Central Axis Commander in the BM forces in Sirte. IS Emir of Tripolitania Province, Abu Hudhayfa al-Muhajir, summarised the main tactics used in an interview with *Rumiyah* magazine: 'booby-traps, explosive devices, tunnels, and encircling manoeuvres . . . [IS detachments] conducted assaults and excursions [that] penetrated [the BM] positions'.[127] Al-Muhajir's propaganda was not too far from the realities on the ground. IS commanders in Sirte managed to effectively coordinate between several squad- and platoon-sized GFs and SGFs. In some neighbourhoods the GFs/SGFs attacks were spearheaded by SVBIEDs, technicals, tanks or infantry fight-ing vehicles. As Muhammad al-Husan, the commander of 166th Battalion who fought IS in both battles of Sirte, stated:

> They were good at manoeuvres, encirclements, infiltrating and then attacking form behind with their *inghimassiyin* [SGFs] . . . They took us by surprise many times. They planted many bombs in our way and used tens of *mufakhkhakhat* [SVBIEDs] . . . constantly counterat-tacking . . . this is how they fought us.[128]

Table 4.2 (overleaf) outlines the categories of tactics reported by IS in Sirte and recorded in the ISWD database between January and December 2016.[129]

Table 4.2 IS Reported Tactics in Sirte, January to December 2016

Category of Tactics	Reported Usage
IEDs	63
SVBIEDs	36
Sniping	16
Artillery	7
Unguided Rockets	7
Guerrilla Formations (GFs)	7
Suicidal Guerrilla Formations (SGFs)	5
MANPADS	4
Anti-aircraft Autocannons	4
ATGMs	3
Close-quarter Assassinations	3
HBIEDs	1
Total	**156**

As expected, IS in Sirte relied heavily on static IEDs, SVBIEDs and sniping tactics during fighting. The organisation used other tactics as outlined above, in addition to armoured and tunnel warfare.[130] The last two categories were mainly used in the south of Sirte (around the Ouagadougou Conference Centre) and in the northeast coastal areas such al-Jizah area.

Regarding static IEDs, the organisation tried to use them in longer-range strikes. The attempts were only party successful, however. As Colonel Ismail Shukri of the Libyan military intelligence stated:

> We [BM forces] found bombs [IEDs] wires several kilometres long. In one operation, they connected the wires with 44 heavy [155mm] artillery shells to blow up a [BM] column. It did not explode. Can you imagine the ramifications if it did?[131]

IS in Libya relied significantly on artillery shells, anti-tank mines, unguided air-to-ground bombs and warheads of surface-to-air missiles[132] as IEDs as opposed to manufacturing IEDs like ISI had done in Iraq.

Regarding SVBIEDs, the heaviest two waves struck the BM forces in May 2016 (12 SVBIED- and SLVBIED-led attacks) and September 2016 (8 SVBIED-led attacks) after the American intervention.[133] 'It was a suicide car after a suicide car or what they [IS units] call 'dogma' . . . followed by infantry formations firing intensely', said al-Siddiq al-Soor, the Director of Investigations in Libya's Attorney General's Office.[134] Armoured SVBIEDs and SLVBIEDs were particularly used to initiate offensives or stave off the BM advances. 'They [IS Units] also attacked with large-truck carrying 20 tons of explosives', said Colonel Shukri,[135] going on to say, 'we called them "death trucks" [SLVBIEDs] they carried aerial bombs . . . they [IS units] took them [air-to-surface bombs] from the ammunition depots in Khushum al-Khayl area [90 miles south of Sirte]'.[136] IS in Sirte also used drones to record SVBIED-led attacks during the battle of Sirte. It is unclear whether the drones were utilised to film and guide the SVBIED – as was the case in the battle of Mosul[137] – or to only film the SVBIED operations for propaganda purposes. Overall, the SVBIEDs and SLVBIEDs were the most effective weapon in IS arsenal, particularly in Sirte. Before the American close air-support in August 2016, the BM forces' responses to SVBIEDs attacks had a limited effect. 'We could not do much. We just fired on [fast-approaching] boobytrapped cars with light weapons. It still killed many of our soldiers and wounded more', said Colonel Shukri.[138] Finally, IS focused the use of static VBIEDs in urban terrorism operation in Misrata, Benghazi and smaller cities and towns (such as al-Qubba). The battles of Derna and Sirte did not witness any recorded waves or significant numbers of VBIEDs.

The future of IS insurgency in Libya

IS in Libya has shown a rapid ability to build-up its military capabilities via the iALLTR *modus operandi*. During its battles in Derna and Sirte, the organisation strategically shifted between conventional, guerrilla and terrorism ways of warfare to avoid annihilation. From late 2014 onwards, the organisation was able to operationally and tactically innovate to pull off multiple upsets against relatively stronger enemies. Many of these innovations arrived in Libya via 'knowhow-transfer' from either Iraqi commanders or veterans of combat zones abroad. These veterans were both Libyans and non-Libyans.[139] The innovations included modified and upgraded urban terrorism, guerrilla warfare and conventional military tactics imported from IS Provinces in Iraq and Syria and their predecessors, namely ISI and ISIS. Covert and static IEDs, SVBIEDs and sniping tactics were crucial for the organisation's military victories in Libya and the abundance of military munitions such as surface-to-air missile warheads, air-to-surface bombs, artillery shells and anti-tank mines enabled IS to utilise them as the explosive-load part of the IEDs.[140]

After the defeats in Derna, Sirte and other Libyan towns, IS in Libya re-configured its strategy to guerrilla warfare and urban terrorism. That shift meant dissolving existing structures and dispersing forces into the southern desert as small guerrilla bands and in larger cities as small terror cells. In February 2018, fourteen months after IS defeat in Sirte, Cyrenaica Province declared in al-Naba' newsletter the resumption of a guerrilla warfare campaign (war of attrition) across Libya.[141] IS initiated the campaign with two SVBIED attacks and one GF-led attack, all targeting Hefter's forces in eastern Libya. In July 2018, IS reorganised and merged all of its 'provinces' in Libya (and elsewhere) in response to territorial losses and a series of tactical and operational defeats. In Libya, the three provinces of Cyrenaica, Tripolitania and Fezzan were merged into 'Libya Province'. The newly formed organisational structure initiated its campaign with urban terrorism tactics in both western and eastern Libya. In December 2018, a small inghimassiyun (SGF) unit of three fighters attacked and burned down parts of the Libyan Ministry of Foreign Affairs in Tripoli. Al-Naba' newsletter called the SGF unit a 'Libya Province security detachment'.[142] In June 2019, IS struck back in Derna with two parked VBIEDs targeting Hefter's LNA forces, after they occupied the town and defeated Derna Protection Force.[143]

In addition to urban terrorism, the new IS structure was also focused on building guerrilla capacities and reigniting a rural guerrilla warfare campaign in the large, sparsely populated and strategically significant south of Libya (Fezzan). 'Unfortunately, the Libyan [southern] desert is still full of IS forces, so we have to be very careful and stay alert, defending the borders of Sirte', said Brigadier-General al-Ghusri in January 2019.[144] The former spokesperson of the 2016 BM Operation stressed that the GNA's forces had no resources to pursue IS into the southern desert after the liberation of Sirte.[145] Adding to both national and international concerns, al-Naba' newsletter published an interview with the IS military commander of Fezzan on 15 May 2019.[146] Titled 'We will Continue

Storming Towns and Cities', the commander outlined some of the features of the IS guerrilla warfare campaign in Fezzan. They include temporal control of towns and villages via storming them, looting available resources, liquidating informants and collaborators, terrorising potential informants and collaborators by burning their homes and confiscating their properties and then disappearing into the desert.[147] The towns of Ghadwa, al-Fuqaha' and villages near Sabha became victims of that campaign.[148] The IS commander was not bluffing.

In the same issue and in the three previous ones, *al-Naba'* newsletter published a four-part series of articles titled 'Temporal Conquering of Cities as a *Modus Operandi* for the *Mujahidin* [Holy Fighters]'. The articles outlined an operational-level shift to temporarily controlling towns and small cities for various reasons, including looting resources, liquidating collaborators, ambushing isolated posts of anti-IS forces and boosting the morale of IS fighters. It is a modified SCCLC, without the last two 'Cs': no coalition-building and no consolidation. The raids on the desert towns and villages of southern Libya were used to illustrate that *modus operandi*.

Two United Nations Security Council reports issued in January and July 2019[149] concluded that the IS Libya Province increasingly resorts to 'hit-and-run operations out of several points of concentration in Sabha and Jufrah Governorates [southern Libya]'. The January report also stated that IS 'frequently raided and held inner-town police stations in shows of strength and to secure arms. This tactic was repeated in Uqayla', Zlitan, Fuqaha' and Tazirbu'.[150]

At a macro-level, the future of IS in Libya is likely to be affected by the destruction and uninhabitability of major parts of Sirte, Derna, Benghazi and other towns, mainly as a result of the latest phase of the Libyan civil war initiated by Khalifa Hefter's two coup attempts in February and May 2014 and the attempt on Tripoli in April 2019.[151] A United Nations Security Council Report issued in July 2019 observed that IS's

> activity in the south of Libya gained momentum as a consequence of the preoccupation of the Libyan National Army with the battle around Tripoli. Since fighting around the capital began on 4 April 2019, ISIL [IS] fighters have repeatedly attacked the cities of Zillah, Fuqaha' and Fazzan [sic]. They were able to hold these cities for hours at a time and succeeded in freeing some ISIL [IS] prisoners.[152]

IS in Libya and potential like-minded successors are more likely to capitalise on such an environment to endure and, perhaps, expand in the future.

Notes

1. 'Impenetrable Wall' is one possible translation of the Quranic term *al-Bunyan al-Marsus*. The term has poetic and connotative meanings. It is the codename of the operation to liberate Sirte from IS. The operating forces were a coalition of military formations, mainly from the City of Misrata, loyal to the UN-backed Government of National Accord (GNA) in Tripoli.
2. IS used the same Ottoman and Italian political-administrative delineations/subdivisions of Libya. Under the Ottomans and initially under the Italians, the country was composed of three provinces: Cyrenaica (east), Tripolitania (northwest) and Fezzan (southwest). Eric Schmitt and David D. Kirkpatrick, 'Islamic State Sprouting Limbs Beyond Its Base', *New York Times*, 14/2/2015, Accessed on 11/12/2019, at: https://nyti.ms/2LJ4lGJ
3. Martin Pengelly and Chris Stephen, 'Islamic State leader in Libya "Killed in US Airstrike"', *The Guardian*, 14/11/2015, Accessed on 11/12/2019, at: https://bit.ly/2rzn3JR
4. Derna is a coastal city which borders a series of hills called the Green Mountains. Most of the fighting however occurred in urban and suburban areas and not in the Green Mountain. Sirte's geography was slightly helpful for insurgents from all sides. A coastal city in the middle of Libya, it had a strategic depth in the form of the desert to its south which could be accessed via a highway. For more details, see the section entitled 'The Battlefront of Derna' and 'The Battlefront of Sirte' below.
5. IS did not meet significant resistance when it controlled a few buildings and positioned its fighters in Derna's city centre in October 2014. The battle for Derna raged later in June 2015, after series of skirmishes and assassinations.
6. The timeframe usually begins when IS takes over a targeted town/city or parts of it. It ends when IS loses the town/city or the parts it previously occupied.
7. For comparison, see the chapters on IS in Syria and Egypt, as well as the concluding chapter.
8. Some of the newly recruited IS fighters in Libya were defectors from local or foreign regular armies, or experienced fighters from other organisations. This is in addition to young men and teenagers with low-to-no experience.
9. Some of these fighters were already founders of the IYSC. Noman Benotman. Former Commander in the Libyan Islamic Fighting Group, Interview by author, Prague, 16 August 2019.
10. Security Council, United Nation, 'Security Council ISIL (Da'esh) and Al-Qaida Sanctions Committee Adds 12 Names to Its Sanctions List', *Press*

Release, SC/12266, 29/2/2016, Accessed on 11/12/2019, at: https://bit. ly/38pvTKM

11. Ibid; Benotman, Interview by author.

12. 'Statement from Pentagon Press Secretary Peter Cook on Nov. 13 airstrike in Libya', *US Department of Defense*, Immediate Release, 7/12/2015, Accessed on 11/12/2019, at: https://bit.ly/2qGBN9r

13. Aidan Lewis, 'New Islamic State leader in Libya Says Group "Stronger Every Day"', *Reuters*, 10/3/2016, Accessed on 11/12/2019, at: https:// reut.rs/3566jbH

14. See: 'Eulogy to Abu Nabil al-Anbari: Islamic State leader in Libya', *aymennjawad.org*, 7/1/2016, Accessed on 11/12/2019, at: https://bit.ly/2RMdzG1

15. Paul Cruickshank, Nic Robertson, Tim Lister and Jomana Karadsheh, 'ISIS Comes to Libya', *CNN*, 18/11/2014, Accessed on 11/12/2019, at: https:// cnn.it/2RDFVBZ

16. Abdul Hakim Belhaj, Former Commander (Emir) of the Libyan Islamic Fighting Group (LIFG), Interview by author, Doha, 27 January 2019.

17. Michelle Nichols, 'Islamic State in Libya Hampered by Lack of Fighters -U.N. Experts', *Reuters*, 1/12/2015, Accessed on 11/12/2019, at: https:// bit.ly/2LXXwkL

18. IS fighters remained in the outskirts of Derna, particularly in al-Fattaih area, until their complete retreat to Sirte in April 2016.

19. A Coalition of mainly Misratan military formations, loyal to the GNA. *Al-hunyan al-marsus* is an Arabic Quranic term, with poetic and connotative meanings (Chapter *al-Saff*, 61:4). It can be translated as 'Impenetrable Wall' or 'Solid Foundation'.

20. Julian Hattem, 'CIA: Undaunted ISIS is Expanding, Focused on Attacking West', *The Hill*, 16/6/2016, Accessed on 11/12/2019, at: https://bit. ly/2E6Jd91

21. Jared Malsin and Benoit Faucon, 'Islamic State's Deadly Return in Libya Imperils Oil Output', *Wall Street Journal*, 18/9/2018.

22. Maggie Michael, 'How a Libyan City Joined the Islamic State group', *AP News*, 9/11/2014, Accessed on 11/12/2019, at: https://bit.ly/2t4xnKd

23. The Arabic acronym for 'Islamic State in Iraq and Sham' (ISIS). *Daesh* is usually used as a derogative term to describe IS and its predecessors.

24. Abdelwahab al-Haddad and Abu Bakr Al Dharrat, 'The Confessions of "ISIS Dinosaur" . . . The Princes' Driver Reveals the Leaders of the Organization and Smuggling Routes to Syria' [In Arabic], *Al-Araby al-Jadid*, 18/11/2018, Accessed on 11/12/2019, at: https://bit.ly/2Ecx0iV

25. Ibid.

26. Islamic State Ways of Warfare Dataset in Derna 2016 (ISWD-Derna16); these figures are a part of the 'Islamic State Ways of Warfare' Database or ISWD, based on IS declarations in *al-Naba'* newsletter between issue no. 12 and issue no. 27. These *al-Naba'* issues were published during the Battle of Derna. IS units completely withdraw from Dern as a guerrilla force on 20 April 2016.

27. 'The Islamic State Media Office in Tripoli Provides The Battle of Sheikh Abu Anas al-Libi, may God Accept Him', [In Arabic] *justpaste.it*, 5/4/2015, Accessed on 11/12/2019, at: https://bit.ly/359afIE

28. Hany Yacine, 'Would Libya Embrace ISIS?' [In Arabic] *Al Ghad TV*, 5/102017, Accessed on 11/12/2019, at: https://bit.ly/2RF7iM7

29. Sudarsan Raghavan, 'Even with US Airstrikes, a Struggle to Oust ISIS from Libyan Stronghold', *The Washington Post*, 7/8/ 2016, Accessed on 11/12/2019, at: https://wapo.st/36ip5g6

30. IS calls these armed groups *sahwat* (awakenings). It does so in a derogatory way to delegitimate any anti-IS resistance. The term here is equivalent to 'traitor' or 'mercenary'. It has been commonly used by IS outside of Iraq, including in Syria, Libya, Egypt and elsewhere.

31. The pictures of the converted infantry fighting vehicle was issued by *Wilayat Barqa* (IS Cyrenaica Province) in a photographic report covering the battle codenamed by the DMSC as 'The Martyrs of al-Qubba' (see details in the next section).

32. For simplification, the DMSC will stand for the shortened title of 'Derna Mujahidin Shura Council' throughout the book. The DMSC also included fighters from foreign Jihadist organisations hostile to IS, including leaders and individuals from the Egyptian *al-Murabitun* organisation; a small armed organ-isation led by former Egyptian Special Forces Major, Hisham al-'Ashmawy.

33. 'Meeting of al-Bayda Tribal Elders with Ali Hassan al-Jaber Brigade's Com-mander in Fattaih-Derna', *YouTube*, 29/9/2015, Accessed on 11/12/2019, at: https://bit.ly/359wpKN

34. 'Mujahideen Shura Council' in Derna declares war on "ISIS"', *Assabeel*, 11/6/2015, Accessed on 11/12/2019, at: https://bit.ly/2YEtB5T

35. The self-declared 'Libyan National Army' (LNA) is a hybrid armed forma-tion of tribal and regional militias and some of the regular units in the former Libyan army under Gaddafi. The 'LNA' was founded by Khalifa Hefter and it is primarily supported by the UAE and Egypt (and later on by France, Jordan and Russia). For more details see: Mary Fitzgerald et al., 'A Quick Guide to Libya's Main Players', *European Council of Foreign Relations* (December 2016), Accessed on 11/12/2019, at: https://bit.ly/35eXix2; Fitzgerald, 'Libya's Rogue 'War on Terror', *Foreign Policy*, 5/6/2014, Accessed on 11/12/2019, at: https://bit.ly/34cZTWS;

36. Sami al-Saadi, Former Commander in the Libyan Islamic Fighting Group (LIFG), Interview by author, Istanbul, 3 November 2016. The NTC was the *de facto* executive body governing Libya during and after the revolution that overthrew the Gaddafi regime. It ruled parts of Libya from 5 March 2011 to 8 August 2012, when it peacefully handed over authority to the first-ever freely elected Libyan parliament. It was the first time that a peace-ful transfer of power had occurred in Libya's modern history.

37. This was a part of a failed reintegration attempt of armed organisations into formal armed institutions. The saga of failed demobilisation and reintegration in Derna and Eastern Libya has been remarkable, with tragic ramifications.

This happened despite the enormous amount of funding and effort dedicated to it. It was mainly spoiled due to local warlordism, interventions of regional authoritarian regimes mixed with high levels of corruption and incompetence. This was one of the main macro-level causes behind the rise of IS in Libya. For more details, see: Omar Ashour, 'Between ISIS and a Failed State: The Saga of Libyan Islamists', in Shadi Hamid and William McCants (eds.), *Rethinking Political Islam* (Oxford: Oxford University Press, 2017), pp. 101–19; Frederic Wehrey, *The Burning Shores: Inside the Battle for the New Libya* (New York: Farrar, Straus and Giroux, 2018).

38. See: 'Interview with Abdul Qadr al-Najdi, the Emir of the Islamic State in Libya', [In Arabic] *al-Naba'* newsletter, no. 21, 8/3/2016, pp. 8–9.

39. Al-Saadi, Interview by author.

40. Like many Wahhabi individuals and entities, the IYSC and IS believe that democracy is a 'belief system' and a 'religion'; as opposed to a system of governance and an executive conflict resolution mechanism for diverse societies. See for example the series of lectures entitled '*al-dimuqratiya din*' [democracy is a religion] by Abu Muhammad al-Adnani, the former spokesman of IS. See also: IS *Dabiq* Magazine, no. 14, pp. 34–8.

41. The title/term has a strong historical significance/connotation in Muslim societies. It refers to a battle in 659 ad fought between Ali ibn Abi Talib, the cousin and son-in-law of the Prophet Muhammad and the fourth caliph, and a group-turned-sect that opposed him called the 'Kharijites' near al-Nahrawan (about 20 miles east of Baghdad). The battle ended in a defeat of the Kharijites. The Kharijites are a minority sect considered by Sunnis and Shiites as 'heretics'. IS (organisation) and the Kharijites (sect) have a few similarities. Both are dogmatists, anti-status-quo and violent. But there are major differences in their belief-systems/ideologies and behaviours. Anti-IS groups and official religious establishment call IS/ISIS/ISI (and other groups) 'Kharijites' to delegitimate them in Sunni-majority societies. The name-calling and comparison are more of a propaganda/counter-propaganda tool. It does not have any comparative, scholarly merit. As a hyper-extremist Sunni group, IS has much more in common with Wahhabi tribal militias of Al-Saud especially during the period between 1902 and 1929/onwards, than any other Sunni or non-Sunni group.

42. The name was given in reference to the victims of three car-bombings perpetrated by IS in the small town of al-Qubba in February 2016. The explosion killed tens of civilians. For more details, see: 'IS Militants Claim Deadly Bombings in Libya', *VOA News*, 20/2/2015, Accessed on 11/12/2019, at: https://bit.ly/2LIcSJH

43. Former officer in Ali Hasan al-Jaber Battalion, Interview by author.

44. Abd al Aziz Besha, 'The Islamic State Withdraws From Derna, Eastern Libya', [In Arabic] *Al Jazeera*, 20/4/2016, Accessed on 11/12/2019, at: https://bit.ly/2RHj6NO

45. Based on *al-Naba'* reports and several photographic reports issued by *Wilayat Barqa* (IS Cyrenaica Province) Media Office covering the battles of Derna.
46. See for example: 'On the Killing of Salim Derby and the Developments in the City of Derna', [In Arabic] *YouTube*, 10/6/2015, Accessed on 11/12/2019, at: https://bit.ly/2sbB9RD; 'Interview with the Former Commander of the Abu Salim Martyrs Brigade: Salim Derby', [In Arabic] Program the Meeting of the Hour, *Libya Al-Ahrar TV*, 24/11/2019 (2013), Accessed on 11/12/2019, at: https://bit.ly/36nN51p
47. See: '*Al-Rimal al-Mutaharika:* Derna', [In Arabic], Documentary, 25/2/2018, Minute 13:10, Accessed on 11/12/2019, at: https://bit.ly/2REamIo
48. Ismail Shukri, commander in the Libyan Military Intelligence branch in Misrata, Interview in: '*Al-Rimal al-Mutaharika*: Derna', [In Arabic] Documentary, 25/2/2018, minute 18:02, Accessed on 11/12/2019, at: https://bit.ly/2REamIo
49. Colonel Muftah Hamza, Commander of al-Hila Frontline against IS in Derna, Interview in: '*Al-Rimal al-Mutaharika*: Derna', [In Arabic] Documentary, 25/2/2018, minutes 24:30–26:40., Accessed on 11/12/2019, at: https://bit.ly/2REamIo
   Colonel Muftah Hamza became the commander of the frontline as of 15 November 2015, after the death of the former commander Colonel Mohammad Boughifayir.
50. IS called the DMSC and other anti-IS units in Derna 'Awakening Councils', like the anti-IS tribal militias in Iraq in 2006–7. This is a form of derogatory propaganda to delegitimate any local anti-IS movements as America's proxies.
51. 'ISIS Admits the Loss of Derna on Video', [In Arabic] *Arabi 21*, 12/7/2015, Accessed on 11/12/2019, at: https://bit.ly/2LKBhOT; 'ISIS Admits the Loss of Derna', *Afrigatenews.Net*, 12/7/2015, Accessed on 11/12/2019, at: https://bit.ly/2qCx0Wm
52. Martin Pengelly and Chris Stephen, 'Islamic State leader in Libya "Killed in US Airstrike"', *The Guardian*, 14/11/2015, Accessed on 11/12/2019, at: https://bit.ly/2rzn3JR
53. *al-Naba'* newsletter, n. 21, 8/3/2016, p. 2.
54. 'Meeting of Al-Bayda tribal elders'.
55. Hamza, minutes 32:00–33:40.
56. 'The Commander of the Operations Room of al-Fattaih Reveals the Name of ISIS Leader in the Area', [In Arabic], *al Wasat*, 22/3/2016, Accessed on 11/12/2019, at: https://bit.ly/2RIq994
   As opposed to the testimony of Colonel Hamza, Abu Soliman al-Tajuri – a former IS commander in Sirte – said that the majority of IS fighters in Derna were locals. He also stated that the majority of IS fighters in Sirte were non-locals, mainly from other parts of Africa. This is not necessarily a contradiction. Colonel Hamza's remarks were about a specific timeframe (March–April 2016) and specific location (al-Fattaih area). The comments of al-Tajuri were

generic estimates of the fighters' background in Derna (2014–16) and in Sirte (2015–16).
57. See: 'Interview with Abdul Qadar al-Najdi''.
58. Hamza, minutes 35:00–36:30.
59. Former commander in Ali Hasan al-Jaber Battalion (Bayda), Interview by author, Tunis, 23 September 2017.
60. Brigadier-General Muhammad Al-Ghusri. Spokesperson of *al-Bunyan al-Marsus* (Impenetrable Wall) Operation, Interview in: '*Al-Rimal al-Mutaharika*: Derna', [In Arabic] Documentary, 25/2/2018, minutes 37:00–38:30, Accessed on 11/12/2019, at: https://bit.ly/2REamIo
61. Sirte is the hometown of Mu'ammar al-Gaddafi. It is one of his strongholds, where significant numbers of his tribe (the Qadhadhfa, pronounced Gaddadfa) are located. Like Derna, Sirte is located on the Mediterranean coast.
62. Mu'ammar al-Gaddafi was born near Qasr Abu Hadi, a rural area outside Sirte in Tripolitania (northwestern Libya).
63. al-Gaddafi family came from a small, relatively uninfluential tribal group called the Qadhadhfa (pronounced Gaddadafa).
64. A note should be mentioned here: the popularity of Islamists in general, and Jihadists in particular, in Derna is highly inflated in the media and even in some of the academic literature. This is clearly shown in the results of the first free and fair parliamentary elections in Libya's history in July 2012. In the first electoral district (where Derna is located), the non-Islamist National Forces Coalition (NFC) swept the polls with over 62,000 votes. The Islamist Justice and Construction Party (JCP) of the Muslim Brothers (MB) only received 8,619 votes. The non-Islamist Central National Trend (CNT) came in third with 4,962 votes. Overall, over 85% of the first electoral district voters chose non-Islamist parties.
65. In contrast to the common myth, IS success in the recruitment of ex-Baathists in Iraq was limited. The overwhelming majority of 'Baathists' either re-joined state institutions (military, security and other branches of government) or non-IS/anti-IS guerrilla organisations. For the 'official' IS position on Baath Party loyalists, see for example: 'Dismantling Armed Units of the Baath Party Apostates', [in Arabic] *al-Naba'*, no. 1, Hijri-dated 03/01/1437 (or 17/10/2015), p. 9.
66. See for example: 'Repentance of 42 Elements Working in the Ministry of Interior', [in Arabic] *Wilayat Tarabulus Media Office*, 14/2/2015.
67. General Youssef Mangush, Former Chief of Staff of the Libyan Armed Forces, Interview by author, Istanbul, March 2017.
68. 'Interview with Ahmed Gaddaf al-Dam, Gaddafi's Cousin and one of the Main Leaders and Former Officials of his Regime', Program 10 p.m., *Dream TV*, 17/1/2017, Accessed on 11/12/2019, at: https://bit.ly/2E8FYxW; 'Former Libyan Official Ahmad Qadhaf Al-Dam: I Support ISIS, Which Should Have Been Established 50 Years Ago', *MEMRI TV*, The Middle East Media Research Institute, 17/1/2015, Accessed on 11/12/2019, at: https://bit.ly/2P9NHBZ

69. This is the tribe of Colonel Gaddafi.
70. Abu Soliman al-Tajuri, IS Commander held by the BM Forces, Interview in: '*Al-Rimal al-Mutaharika*: Sirte', [In Arabic] Documentary, 04/06/2017, minutes 06:00–06:30, Accessed on 11/12/2019, at: https://www.youtube.com/watch?v=COlNl9xMghw
71. The site where Gaddafi hosted fellow heads of state and Pan-African conferences.
72. Former Commander in *al-Bunyan al-Marsus* Operation, Interview by author, Istanbul, 12/11/2018.
73. Mohammad al Arabi, 'Libya . . . ISIS Broadcasts Al-Baghadi's Speeches in Sirte', [In Arabic] *Al Arabiya*, 13/2/2015, Accessed on 11/12/2019, at: https://bit.ly/2sib1V6; 'The Battle of Sirte: Hypotheses and Facts', [In Arabic], *Libyaalkhabar*, 11/12/2019, Accessed on 11/12/2019, at: https://bit.ly/2PbnjrF
74. 'The Islamic State is Showing its Force on the Streets of Sirte, Libya and the University Closes its Doors', [In Arabic] *France 24*, 19/2/2015Accessed on 11/12/2019, at: https://bit.ly/2PaUECU; 'An ISIS Military Parade in Sirte, Libya (Video and Photos)', [In Arabic] *Arabi 21*, 19/2/2015, Accessed on 11/12/2019, at: https://bit.ly/35cIm2m
75. Ibid.
76. Former Commander in *al-Bunyan al-Marsus* Operation.
77. Ibid. Also see: 'The Battle of Sirte: Hypotheses and Facts'.
78. Most of the armed formations which fought in the two battles of Sirte hail from the City of Misrata.
79. The first-ever freely elected Libyan parliament that was the main legislative body between August 2012 and August 2014.
80. Mohammed Abdullah, Head of National Salvation Front Party, Interview by author, Istanbul, February 2016.
81. Ibid.
82. 'Gaddafi Soldiers are Fighting Under the Banner of the Islamic State', [In Arabic] *Al Quds Al Arabi*, 17/3/2015, Accessed on 11/12/2019, at: https://bit.ly/34dthMK ; 'Fajr Libya is Launching Airstrikes Against ISIS in Sirte', [In Arabic] *Al Jazeera Mubasher*, 24/4/2015, Accessed on 11/12/2019, at: https://bit.ly/2LFryta; Yasmine Ryan, 'ISIS in Libya: Muammar Gaddafi's Soldiers are Back in the Country and Fighting under the Black Flag of the "Islamic State"', *Independent*, 26/3/2015, Accessed on 11/12/2019, at: https://bit.ly/2EiJh5P
83. 'Gaddafi Soldiers are Fighting Under the Banner of the Islamic State'. As opposed to Derna – where members of the same household fought on different sides – the regional-tribal polarisation (as opposed to the ideological polarisation) cannot be discounted in the March–June 2015 battle of Sirte. Most of the 166th Battalion came from Misrata, whereas many IS fighters were from the Sirte-based tribes and clans or from nearby towns like al-Nufaliya. The campaign of 166th Battalion could have been easily interpreted – without ideological or organisational titles – as an attack on Sirte-based tribes by Misratan forces/clans.

84. Tom Westcott, 'No Ammo to fight IS in Central Libya, Says Military', *Middle East Eye*, 5/6/2015, Accessed on 11/12/2019, at: https://bit. ly/2PwMqnH
85. Ibid.
86. Soliman al-Zawi and David Kirkpatrick, 'Western Officials Alarmed as ISIS Expands Territory in Libya', *New York Times*, 31/5/2015, Accessed on 11/12/2019, at: https://nyti.ms/2sgjmZH
87. Ibid.
88. Former Commander in *al-Bunyan al-Marsus* Operation, Interview by author.
89. Pronounced 'Bugrain' in Libya, it is a town which is about midway between Sirte and Misrata cities. It is located within the Misrata district. It has a strategic location on the crossroads between the Libyan Coastal Highway and the Fezzan Road, thus giving direct road-access to east, west and south of Libya.
90. The term comes from a Quranic verse. It can be translated as 'Impenetrable Wall' or 'Solid Foundation'. The BM coalition included mainly the 166th Battalion and Misratan company- and platoon-sized formations, in addition to other company-sized formations that came from towns such as Zliten and Sabha.
91. Christian Clausen, 'Providing Freedom from Terror: RPAs Help Reclaim Sirte', *Air Combat Command*, 1/8/2017, Accessed on 11/12/2019, at: https://bit.ly/2RFoe59; 'US Airstrikes in Support of the GNA, October 13', *US Africa Command Press Release*, 14/10/2016, Accessed on 11/12/2019, at: https://bit.ly/2PAj6fP; Adam Entous and Missy Ryan, 'In Libya, United States Lays Plans to Hunt down Escaped Islamic State Fighters', *The Washington Post*, 11/11/2016, Accessed on 11/12/2019, at: https://wapo.st/2LKsmNz; Sudarsan Raghavan, 'Even with US Airstrikes, a Struggle to oust ISIS from Libyan Stronghold', *The Washington Post*, 7/8/2016, Accessed on 11/12/2019, at: https://wapo.st/2sfdGir
92. Edward Chang, 'The Harrier: The US Marine Corps Loves This Plane For 1 Big Reason', *The National Interest*, 29/3/2019, Accessed on 11/12/2019, at: https://bit.ly/38si1zk
93. 'Three British ISIS Members Killed by SBS Soldiers in Sirte', *Libyan Express*, 4/7/2016, Accessed on 11/12/2019, at: https://bit.ly/36o9hZd; Miss Ryan. 'US Special Operations Troops Aiding Libyan Forces in Major Battle against Islamic State', *The Washington Post*, 9 August 2018, Accessed on 11/12/2019, at: https://wapo.st/35dQNdR
94. Tom Westcost, 'IS Seizes Libya Airbase after Misrata Forces Pull out', *Middle East Eye*, 30 May 2015, Accessed on 11/12/2019, at: https://bit. ly/35aaraA; Former Commander in *al-Bunyan al-Marsus* Operation.
95. Ibid.
96. Ibid.
97. Abu Hudhayfa al-Muhajir, Interview by Anonymous, *Rumiyah*, no. 4, Hijri-dated Rabi' al-Awal 1438, p. 10

98. Islamic State Ways of Warfare Dataset in Sirte 2016 (ISWD-Sirte16).
99. Ibid.
100. Libya was divided into 22 administrative districts. They are sometimes referred to as 'governorates' or '*shabiyat*' (one of Gaddafi's numerous 'neologisms' that can be translated to 'popularates'). The geographical-administrative divisions are similar to Egypt's *muhafazat* (governorates) system and Tunisia's *wilayat* (provinces) system.
101. Seif al-Din al-Trabulsi, 'Daesh Bombing Kills 30 Soldiers Near Libya's Sirte', *Anadolu Agency*, 5/19/2016, Accessed on 11/12/2019, at: https://bit.ly/35np9Lm
102. Loyalties were rapidly shifting in Libya. The same militia could be 'loyal' to opposing sides in different timeframes. Ideological, clan-tribal, regional, political and economic factors determined varying allegiances; with the last two factors specifically changing rapidly.
103. Ayman al-Warfalli, 'Libyan Security Forces Pushing Islamic State back from Vicinity of Oil Terminals', *Reuters*, 30/5/2016, Accessed on 11/12/2019, at: https://reut.rs/2LKPefJ
104. Maggie Michael, 'ISIS Militants Retreat from Libya Bastion as Militias Advance', *Military Times*, 9/6/2016, Accessed on 11/12/2019, at: https://bit.ly/2RFIYK1
105. Maggie Michael, '16 Libyan Militiamen Killed in 2 IS Attacks Near Sirte', *Fox News*, 17/6/2016, Accessed on 11/12/2019, at: https://fxn.ws/36m0Yx8
106. Former Commander in *al-Bunyan al-Marsus* Operation, Interview by author.
107. Ibid.
108. Former Commander in *al-Bunyan al-Marsus* Operation, Interview by author.
109. Ibid.
110. Tom Westcott, 'No Ammo to fight IS in Central Libya, Says Military', *Middle East Eye*, 5/6/2015, Accessed on 11/12/2019, at: https://bit.ly/2PwMqnH.
111. Ismail Shukri, commander in the Libyan Military Intelligence branch in Misrata, Interview in: '*Al-Rimal al-Mutaharika*: Sirte', [In Arabic] Documentary, 04/06/2017, Accessed on 11/12/2019, at: https://www.youtube.com/watch?v=COlNl9xMghw
112. Ryan, 'US Special Operations troops aiding Libyan forces'.
113. 'Libyans 'oust' So-Called IS from Sirte Headquarters', *BBC*, 10/8/2016, Accessed on 11/12/2019, at: https://bbc.in/2LFxFxC; Mohammed Abdullah, Head of National Salvation Front Party, Phone Interview by author, 15 February 2017.
114. 'Sirte Battle against ISIS is Entering its Final Stages', [In Arabic] *The New Arab*, 12/8/2016, Accessed on 11/12/2019, at: https://bit.ly/38rhUnR
115. 'AFRICOM concludes Operation Odyssey Lightning', *US Africa Command*, press release, 20/12/2016, Accessed on 11/12/2019, at: https://bit.ly/2ryJC1a

116. Former Commander in *al-Bunyan al-Marsus* Operation, Interview by the author.
117. There are a few exceptions, such as the abovementioned case of al-Nufaliya battle near Sirte, where IS outnumbered its enemies. Also, in Benghazi, IS was massively – not just slightly – outnumbered. It fought in a tactical, short-lived coalition and managed to survive longer than its numbers would suggest. At several points, it even reached compromises with Hefter's LNA similar to the one reached with the SDF on 15 October 2017 during the Raqqa Battle.
118. Clearly – unlike the classic German *blitzkrieg* – IS had no airpower and no dense formations of armoured corps. IS technicals performed the combined role of mechanised and motorised infantry in regular armies. IS manpower, however, was too limited to occupy and control all of Derna.
119. For more details, see the previous chapter on how IS/ISIS fights in Syria.
120. Like Mosul and Fallujah, the 'softening' element included selective targeting of potentially hostile individuals, and the 'creeping' included taking over of critical positions/junctures in the town-centre.
121. This dataset is a part of the 'Islamic State Ways of Warfare' Database (ISWD-Derna16).
122. Ibid. The sample of these categories of tactics are based on the reports issued by *al-Naba'* newsletter between 5 January 2016 to 29 December 2016 (*al-Naba'* issue numbers 12 to 61).
123. It is unclear if that strategic shift was due to the failure in Derna, or to other contextual factors.
124. 'Repentance of 42 Elements Working in the Ministry of Interior', [in Arabic] Photo Report issued by *Wilayat Tarabulus* Media Office, 14/2/2015.
125. After the retreat from al-Fattaih on the late night of 19 April 2016, all remaining IS fighters (about 217) joined IS in Sirte.
126. IS publications did not give any details on the type of MANPADs used. However, based on interviews with BM soldiers and open-source material, it was more likely an SA-16 or SA-18 variant. These were widely available in Gaddafi's army weapon depots. Also, IS had acquired SA-7 missiles from the 136th Barracks near Sirte. However, these missiles were likely dysfunctional due to the lack of parts or due to maintenance issues. For more details see for example: Jonathan Broder, 'Isis in Libya: How Muhammar Gaddafi's Anti-aircraft Missiles are Falling into the Jihadists' Hands', *The Independent*, 11/3/2016, Accessed on 11/12/2019, at: https://bit.ly/35de27R
127. Abu Hudhayfa al-Muhajir, Interview by Anonymous, *Rumiyah*, n. 12, December 2016, p. 10.
128. Muhammad Al-Husan, Commander in the 166th Battalion, Interview in: '*Al-Rimal al-Mutaharika*: Sirte', [In Arabic], Documentary, 04/06/2017, Accessed on 11/12/2019, at: https://www.youtube.com/watch?v=COlNl9xMghw

129. Islamic State Ways of Warfare Dataset in Sirte 2016 (ISWD-Sirte16).
130. There was less reporting on the details of the armoured and tunnel warfare in IS media outlets. However, the usage of tunnel and multiple tanks was verified by other sources involved in the BM Operation. Former Commander in *al-Bunyan al-Marsus* Operation.
131. Colonel Ismail Shukri. See Interview in *Rimal Mutaharika:* Sirte [In Arabic], minutes 27:30–27:58.
132. This includes the warhead of SA-2 missiles, commonly available in the air-defence depots of Gaddafi's army.
133. Islamic State Ways of Warfare Dataset in Sirte 2016 (ISWD-Sirte16). This is based on attacks published in the *al-Naba'*, no. 31 to 37 and no. 46 (September 2016).
134. Al-Siddiq al-Soor, The Director of Investigations in Libya's Attorney General's Office, Interview in: '*Al-Rimal al-Mutaharika*: Sirte', [In Arabic], Documentary, 04/06/2017, minutes 26:28–27:20, Accessed on 11/12/2019, at: https://www.youtube.com/watch?v=COlNl9xMghw
135. See: Interview Colonel Shukri, *Rimal Mutaharika:* Sirte [In Arabic], minutes 27:20–28:30.
136. Ibid.
137. For more details, see Chapter 1 on Iraq.
138. See: Interview Colonel Shukri, *Rimal Mutaharika:* Sirte [In Arabic], minutes 27:20–28:30.
139. In an in-depth study, Aaron Zelin traced 625 Tunisian fighters in Libya via open-source materials, some of whom were IS commanders or fighters. See: Aaron Zelin, 'The Others: Foreign Fighters in Libya''. *Policy Notes of Washington Institute for Near East Policy* (2008), p. 2, Accessed on 20/02/2019, at: https://www.washingtoninstitute.org/uploads/PolicyNote45-Zelin.pdf. See also: Aaron Zelin, *Your Sons are At Your Service* (New York: Columbian University Press, 2020), pp. 242–4.
140. Ibid. The common IED is usually composed of five parts: a power-source, a switch/trigger, a detonator, an explosive load and a container. IS and its predecessors used tens of types of switches, detonators and explosive loads (including air-to-air missiles), in addition to various types of penetrators.
141. *Wilayat Barqa*, 'Caliphate Soldiers Start a New War of Attrition Campaign', [in Arabic] *al-Naba'* newsletter, no. 120, 22/2/2018, p. 6.
142. In Arabic *mafariz amniya*. See: *Wilayat* Libya. 'Soldiers of the Caliphate in Libya Burn Down the Foreign Affairs Ministry', [in Arabic] *al-Naba'* newsletter, no. 162, Hijri-dated 20/06/1440, p. 4.
143. Derna was taken over by Hefter's LNA from the DMSC. The latter reorganised and transformed with its allies into 'Derna Protection Force' on 11 May 2018. See the IS claim of the attack in *al-*Naba', no. 185, Hijri-dated 03/10/1440, p. 8.
144. Tom Westcott, 'No Ammo to fight IS in Central Libya, Says Military', *Middle East Eye*, 5/6/2015, Accessed on 11/12/2019, at: https://bit.ly/2PwMqnH

145. Ibid.
146. *Wilayat* Libya, 'The Commander of the Soldiers of the Caliphate in Fezzan: We Shall Continue Storming Towns and Villages', [In Arabic] *al-Naba'* newsletter, no. 182, Hijri-dated 11/09/1440, p. 7.
147. Ibid.
148. That guerrilla warfare campaign of 'Libya Province' has operational features similar to how Sinai Province in Egypt operate in Sinaian towns and villages.
149. 'United Nations Security Council report', S/2019/570, *United Nations Security Council*, 15/7/2019, Accessed on 11/12/2019, at: https://bit.ly/2qH5eYZ
150. 'ISIL (Da'esh) and Al-Qaida Sanctions Committee: Monitoring Team's Twenty-Third Report', p. 10; 'United Nations Security Council report'.
151. Anas El-Gomati. 'Haftar Rebranded Coups'. *Sada*, 30/07/2019, accessed on 14/08/2019, at: https://carnegieendowment.org/sada/79579
152. Ibid, p. 9.

5

Lures and Endures: How 'Sinai Province' Fights in Egypt

The operations in Sinai are over . . . and we could have finished
terrorism there in six hours.

General Ahmed Wasfy, Commander of the Egyptian Second
Field-Army, 2 October 2013[1]

About 30 Officers and Soldiers are Dead or Wounded in Attacks by
the Islamic State Soldiers in Sinai

Headline of IS *al-Naba'* newsletter issue no. 221,
14 February 2020[2]

The tough weakling

'Don't send reinforcements to Sinai [Sisi]. Send your whole army. It
will die here in the desert', said Kamal Allam, one of the military commanders of *Wilayat Sayna'* or Sinai Province (SP), the most combat-
effective insurgent organisation in Egypt's modern history. Allam's
words were conveyed with a destroyed tracked armoured vehicle
visible in the background.[3] This message was broadcast after simultaneous attacks on seven military and security targets in April 2015.
Two months later, SP launched an even more complex operation.
In July 2015, it attacked 21 military and security targets and briefly
occupied the town of Sheikh Zuweid. Back then, the insurgents used
surface-to-air guided missiles, light artillery, heavy machine guns and
unguided Grad rockets.[4] During fighting, SP insurgents temporarily
cut off targeted posts from incoming reinforcements by using a combination of IEDs, snipers and light artillery. As ever, the number of
dead army soldiers and officers was contested. The military claimed
only seventeen, while unofficial sources claimed over one hundred.[5]
In September 2015, four American and two Fijian peacekeepers were

wounded in two IED blasts near the North Camp of the Multinational Force of Observers (MFO) in Northeastern Sinai. This was one of the relatively few attacks on the MFO since 2005.[6] In October 2015, SP infiltrated Sharm al-Sheikh International Airport in Southern Sinai. SP loyalists planted an IED on the Russian Metrojet Flight no. 9268. The bombing killed all 224 passengers and crewmembers. It was the deadliest terrorist attack yet on Egyptian territories,[7] and also in the history of Russian aviation.

In the first six months of the following year (2016), SP claimed 271 attacks in North Sinai Governorate alone.[8] It intensified its operations with an average of 45 attacks per month in the first half of 2016. This, however, decreased to an average of 40 attacks per month for the whole year, compared with an average of 25 attacks in 2015.[9] Overall, between November 2014, the date of SP foundation, and November 2016, the date of its 2016 annual military report, the organisation claimed more than 700 attacks.[10] SP and its predecessor, *Ansar Bayt al-Maqdis* (Supporters of Jerusalem or ABM), remain the only local insurgent organisations in Egypt's modern history that have been able to acquire quasi-conventional military capacity. This is reflected in the downing of a Second Field Army Mi-17 helicopter in January 2014,[11] the bombing of a naval frigate in November 2015[12] and the capturing of an army tank (documented in an SP video broadcast in August 2016).

The endurance of the SP is perhaps the most puzzling, compared to all other IS Provinces discussed in the previous chapters; not only because SP has even more limited resources, but also because they fight a larger and a more centralised local military. Putting aside its historically mediocre combat-performance, that military is currently funded, trained and supported by a superpower (the United States).[13] This is in addition to tactical, operational, political and other forms of support from a regional power (Israel) and similar forms of support from two wealthy states (Saudi Arabia and the United Arab Emirates).

Given its structural weaknesses, SP's military build-up and survival strategies slightly deviated from the previously analysed IS Provinces in Iraq, Syria and Libya.[14] Despite these structural weaknesses, the Sinaian insurgency was able to endure. At certain timeframes, it was also able to expand its geographical scope, tactical military capacity, operational intensity and durations, regional scale, quality of propaganda and communications and existential legitimacy. This had happened while SP and its predecessors were engaged in combat with unrestrained, superior forces since September 2013.[15]

The endurance and expansion of SP is puzzling for other reasons as well. First, geographically, Sinai's northeastern coastal terrain is not rugged. Most of the high mountains, such as the peaks of Mount Catherine (8,668 feet) and Mount Sinai (7,497 feet), are in the south of the peninsula, far away from the operational focus of the insurgency. Clashes occurred on Halal Mountain (5,577 feet) in central Sinai, but it was not the main theatre of operation. Second, Sinai's population is relatively small. North Sinai Governorate has a population of about 434,781 (40 persons per square mile). Most of the armed action happened in three out of its six districts: al-Arish, Sheikh Zuweid and Rafah (all mainly flat coastal districts on the Mediterranean), with a combined population of about 300,000. Third, loyalty is divided among its population. Almost every northeastern tribe and clan has members and supporters of the insurgency, as well as informants and pro-incumbent tribal militiamen. The divisions do not follow clear rural–urban, settler–Bedouin, or tribal–administrative delineations. Each of these categories has elements on both sides.[16] Finally, there is no state sponsorship for the insurgents. None of the regional governments is directly supportive of the insurgency, including the Hamas authorities in Gaza.[17] On the other hand, the regime's military forces alone enjoy an estimated 100-to-1 overall manpower ratio, at least.[18] This is in addition to the support provided by the United States in terms of financing, training, equipping and intelligence, as well as the support provided by Israel in terms of intelligence-sharing and tactical/operational support. Considering all that, the main two questions of this chapter are how did the SP fight and how did it endure so long and, at certain times, expand?

The military-making of Sinai Province

The saga of the Sinaian insurgency and its military build-up can be traced back more than fifteen years, long before the official establishment of SP and its predecessor ABM. The insurgency, predominantly under ABM and SP, has, however, significantly developed and mutated. Its stated goal changed from supporting Palestinian armed organisations in the early 2000s to controlling areas in northeast Sinai and attempting to defeat the Egyptian regime's security and military forces in the region while declaring transnational loyalties (namely, to

IS in November 2014). Additionally, the insurgency fought to avenge the deaths that occurred during the military crackdown in the aftermath of the July 2013 coup against former President Mohammed Morsi, as well as to weaken the regimes' forces in areas other than the Sinaian Peninsula.[19]

The political, social, security and humanitarian crises that created a hospitable environment for the insurgency go back to the aftermath of the Israeli withdrawal in 1982. Security and social policies since then have essentially branded Sinai as a threat rather than an opportunity, and the Sinaian as a potential informant, potential terrorist, potential spy and/or potential smuggler rather than an Egyptian citizen. In a cable published by *Wikileaks*, a senior Egyptian security official in Sinai told a visiting American official delegation that 'the only good Bedouin in Sinai was the dead Bedouin'.[20] The policies based on this perception escalated after the second Palestinian *Intifada* (uprising) in 2000. At that time, several Egyptian security bureaucracies (principally State Security Investigations [SSI], which is now renamed the National Security Apparatus, as well as the General Intelligence Directorate) believed that there was direct logistical support coming from northeast Sinai to several Palestinian militant groups in Gaza. Since then, the main consistent feature of the security-led policies was a mix of repression and attempted co-option of selected tribal leaders to provide intelligence.

Developing tactics: from urban terrorism to insurgency

In October 2004, the simultaneous bombing of tourist resorts in Taba and Nuweiba marked the beginning of the insurgency and brought about further escalation. The SSI had almost no information about the terrorists and therefore conducted a wide crackdown in Northeast Sinai. With the help of the Central Security Forces (CSF), the SSI arrested around three thousand persons, and took hostage women and children related to suspects until the suspects surrendered. 'They electrocuted us in the genitals for hours before asking any questions. Then the torture continues during and after the interrogations. Many of the young men swore revenge', said one of the former detainees.[21]

In July 2005, a second wave of bombings hit Sharm el-Sheikh. This time, an organisation declared responsibility for the attacks: *Al-Tawhid wa al-Jihad fi Sayna'* (Monotheism and Struggle in Sinai or TJS). The group was inspired by Abu Musab al-Zarqawi's organisation in Iraq, but despite the Iraqi inspiration and the ideological links, most of its leaders and members were locals.[22] The founder, Khaled Musa'id, was a dentist from al-Arish city and a member of al-Sawarka tribe, the largest and one of the most influential tribes in North Sinai. He was killed in a firefight with the CSF on 28 September 2005. Ten years later, in September 2015, SP paid tribute to him as a founder in a 37-minute video entitled 'Soldiers' Harvest.' The main contribution of Musa'id and his men was transforming an ideological current found in books and speeches into an armed organisational structure, with a leadership hierarchy and multiple cells in five cities/towns within three regions: Northeast Sinai (al-Arish, Rafah, Sheikh Zuweid), Central/Central-East Sinai (Halal Mountain/Nekhel) and Ismailia city on the western bank of the Suez Canal.

A second wave of crackdowns began immediately after the 2005 bombings. Many suspected TJS members and sympathisers as well as their relatives, acquaintances and neighbours were arrested. 'We met them in prison. Most of them did not know anything about ideology, theology, or jurisprudence. Some were illiterate, and we had to teach them how to read', said a former Islamist detainee who was imprisoned with the 'Sinai group', as they were known at the time.[23] 'All that the actual TJS members had studied were three booklets written by Abu Muhammad al-Maqdisi', he said, referring to a famous Jordanian Jihadist ideologue, 'and this led them to use *takfir* [excommunication] a lot.'[24]

The Sinai detainees were mainly distributed in six prisons: Damanhur, Highly Guarded (known as the Scorpion), Abu Za'bal, Liman Tora, New Valley and Natrun Valley.[25] From 2004 to 2009, they interacted in those prisons with former Jihadists who had abandoned and disowned armed activism. The SSI allowed the Islamic Group (IG), several former leaders of al-Jihad Organisation and independent Salafi figures, to give them lessons in Islamic jurisprudence (*fiqh*), creed (*'aqida*) and the 'revisions literature' produced by the IG and others in their post-jihadist phase, thinking it would provide a fruitful counternarrative to Jihadism.[26] It partly worked. In prisons, some of the TJS members abandoned the core jihadi belief that armed action is the sole theologically legitimate, instrumentally effective, method for social and political change.[27] Others did not and were more interested in avenging their humiliation and repression. This was notably the case with Kamal Allam, one of SP's military commanders quoted above, as well as Tawfiq Firij Ziada, one of ABM's founder and the alleged architect behind its organisational centralisation.[28]

Back in Sinai, the environment was changing significantly. A 2007 conflict in Gaza between factions of Hamas and Fatah, and a 2009 crackdown by Hamas on Salafi–Jihadists, drove Fatah's Preventive Security officers and members of *Jund Ansar Allah* (Soldiers of the Supporters of God) and their sympathisers into northeast Sinai to escape the crackdowns by Hamas authorities. By late 2009, Jihadists began to regroup in different organisational structures. *Al-Tahwid wa al-Jihad – Bayt al-Maqdis* (Monotheism and Struggle – Jerusalem) issued a booklet on 17 November 2009 entitled 'This is Our Creed'.[29] It did not differ much from the theological booklets issued at that time by the Islamic State in Iraq (ISI) to support its ideological version of Jihadism. The group *Ansar al-Sunnah fi Aknaf Bayt al-Maqdis* (Supporters of the Sunnah near Jerusalem) issued a series of propaganda videos documenting operations against Israeli civilian and military targets between April and August 2010. By late 2010, individuals and factions from these small groups had merged to become ABM, which became the most active and centralised armed organisation in Sinai, among at least three other active groups and networks.[30]

Between late 2010 and late 2013, ABM was focusing primarily on attacking Israeli civilian and military targets. But Egyptian police stations and security headquarters were also attacked in January, February and July 2011, partly to avenge the crackdowns of 2004–6 as well as to loot weapons. Unlike Cairo, Sinai's uprising against Hosni Mubarak's regime combined both popular mobilisation and armed reprisals on security forces. As a result, by early February 2011, security forces had fled both the towns of Rafah and Sheikh Zuweid.

Shifting the aim: targeting another enemy

In response, two major counterinsurgency operations were launched by the Egyptian governments between 2011 and 2013: Eagle One (under the Supreme Council of the Armed Forces, or SCAF) and Eagle Two (under President Mohammed Morsi). Both operations failed to quell the insurgency and, in many ways, seemed to be a continuation of previous inefficient policies, rather than marking a new counterinsurgency approach. The July 2013 coup and the August 2013 massacre of protestors in *Rabaa* and *al-Nahda* Squares in Cairo had a major impact on Sinai. 'We knew torture is on its way again. It was just a matter of time', said one of the residents of Sheikh Zuweid who was detained for a few months in 2004, before being released uncharged.[31] In August 2013, Sinaian *Salafi* figures and preachers held a public conference in Sheikh Zuweid in which one of the speakers demanded the formation of a 'war council' to fend off an expected wave of repression.

After the crackdown in *Rabaa* Square, there was a significant change in the insurgency's rhetoric, behaviour, intensity and scale of operations, as well as in its overall narrative and goals. In a rare interview an ABM member stated that:

> Morsi admitted leading the campaign against us. But [at that time] the reply of our brothers was limited to defence. We did not deliberately attack military headquarters or follow the security officers to target them. But after what happened after the military coup, fighting the armed forces became an urgent necessity.[32]

Indeed, ABM was highly sensitive to the local developments. It changed its rhetoric and narrative primarily to stress the idea that it was 'defending the Muslims of Egypt against the onslaught of an 'Army of Murderers and Apostates'.[33] This was a significant departure from the narratives of 2010–12, when the organisation stressed that it was targeting Israel and its interests, while attempting to avoid a clash with the local military.

Absorption by the Islamic State: the story of the Bay'a

On 22 December 2016, *al-Naba'* newsletter published an interview with the former leader of Sinai Province, the so-called '*wali*' (governor) of Sinai Province, Abu Hajar al-Hashimi. Egyptian authorities claim that the latter is a former Iraqi officer.[34] Regardless, the interview reflected the common IS *modus operandi* (iALLTR) in terms of the absorption of local armed organisations, individual and collective recruitment, leading by experienced battle-hardened locals and foreigners and transference of tactical knowhows and innovations. Abu Hajar highlighted these knowhow transfer and leadership dimensions:

> As for the *modus operandi* (of IED warfare), it remains like Iraq; with some modifications in communication mediums and remote detonation . . . IEDs are the perfect weapon for dealing with the armoured vehicles of the apostate [Egyptian] army.[35]

The ideological and organisational links between the predecessors of SP in Sinai and the predecessors of ISI/ISIS/IS in Iraq go
back twelve years before that interview. The links go beyond the
namesake of the predecessors (*al-Tawhid wa al-Jihad* in Iraq in 2003
[eponym] and *al-Tawhid wa al-Jihad* in Sinai in 2004). By November
2014, IS had successfully absorbed most of ABM's factions into its
Sinai Province, recruited members from the Delta and Nile Valley
regions and established cells there, transferred knowhows to Sinai
from other IS Provinces as well as cadres and leadership, and built up
intelligence capacities that will be used later in 2015.[36] From all five
pillars of the iALLTR *modus operandi*, the organisation came up short
on two: recruitment and looting of large weapon depots.

The November 2014 *bay'a* (religious-political pledge of allegiance) by most of ABM's factions to IS (thus, establishing SP) was
perhaps the most critical and unprecedented development in the history of Egyptian Jihadism. It was the first time that a local armed
jihadist organisation of ABM's size declare a transnational religious
oath of loyalty to a foreign organisation. In late 2013, unconfirmed
reports circulated that Kamal Allam and several other Sinaians had
trained in ISIS camps in Syria.[37] Other reports mentioned that Abu
Bakr al-Baghdadi met with some of the Sinaian Jihadists who fought
in Iraq. He was interested in providing logistical support to them as
early as 2011.[38] In September 2014, IS' official spokesperson, Abu
Muhammed al-Adnani, called on the 'brothers' in Sinai to fight the
regime's army 'in any possible way and to turn their lives into hell
and horror'. Overall, the signs of inclination towards ISIS were clear
from early 2014, despite a speech broadcast on 24 January 2014 by
al-Qaida's leader, Ayman al-Zawahiri, in which he described ABM
as 'our men in Sinai.'

A clear indicator of the successful absorption of ABM by IS came
on 24 October 2014. 'Give the good news to [Abu Baker] al-Baghdadi . . . Give the good news to the caliph of the believers.[39] Victory
is coming. And we are your soldiers, God willing', said a masked
insurgent in a distinct North Sinaian accent on 24 October 2014, after
successfully destroying a military checkpoint in Karm al-Qawadis.[40]
And on 10 November 2014, the affiliation with IS became official.

> The caliphate has been declared in Iraq and al-Sham [parts of the
> Levant], and the Muslims have chosen a caliph who is the grand-
> son of the best of humans. If that is the case, we have no choice

but to heed the invitation of God's caller . . . We therefore pledge religious-political loyalty to Caliph Ibrahim ibn Awad ibn Ibrahim al-Qurayshi al-Husayni.

That month, ABM changed its name to *Wilayat Sayna'* (Sinai Province of the Islamic State, or SP). The statement, which was disseminated by ABM/SP's media section, was probably spoken by the organisation's commander at the time, Abu Osama al-Masri. It put an end to conflicting statements issued by ABM-affiliated media outlets. An earlier statement, issued on 3 November 2014, had declared allegiance to IS and called on the 'brothers in the Land of the Quiver [Egypt], Libya, Gaza and all Maghreb and Mashreq countries' to declare loyalty to IS. But on 4 November 2014, a tweet denying the oath and challenging the authenticity of the first statement was published by one of two Twitter accounts that regularly disseminate ABM's statements. The two conflicting declarations reflected divisions within the organisation. The division was primarily about swearing allegiance to IS, or preserving an undeclared affiliation and ideological affinity with al-Qaida, especially its post–2011 modified strategy of mixing political violence with social services to win popular support.

Why did SP militarily endure?

The previous section showed how SP developed organisationally. This section addresses the main causes of its durability: its military capacities and resources, the counterinsurgency blunders committed by the regime's forces and the socio-political environment in which both sides operate. Other elements matter, but to a much lesser extent. They include, for example, how the narrative and the propaganda of SP bolsters its military efforts and how it affects its legitimacy among segments of the Egyptian and the Sinaian communities. These elements will be referred to, but not focused on here.[41] The section does not address conspiracy theories, 'agent provocateur', 'false flag' and 'limited aims' arguments, mainly due to space limitation. The summary of these arguments is that the military's aim is to contain and weaken the insurgency in the remote periphery but does not aim to destroy it. The reasons behind this limited aim are to maintain international legitimacy, divert attention from gross human rights abuses, and to keep the flow of logistical support, including arms, equipment, funds, and intelligence on the basis that the regime is still at war with a military credible IS affiliate. The regime also manipulates the insurgency, so such arguments go, to eliminate/assassinate factional opposition within its ranks and/or opposition figures outside of it, in a classic 'false flag' tactic. Most of these arguments, however, cannot be established, refuted, or credibly tested. Both 'false flag' and 'agent provocateur' tactics, however, were used by military and security agencies in Egypt many times before. This was established primarily by the memoirs and testimonies of high-ranking military officials, including ministers and a former president.[42] In the case of SP, far less credible testimonies have surfaced to bolster such arguments.[43]

Military capacity

The combat capacity of SP is unprecedented in Egypt's history of insurgencies, including the 1992–7 low-level insurgency in Upper Egypt[44] and the 1952–4 clashes in Greater Cairo.[45] SP mainly employs and shifts between two ways of warfare. The first includes the common tactics of 'urban terrorism'. This includes attacks in cities and towns via a combination of car bombs, suicide attacks and targeted assassinations. The second type is guerrilla warfare; mainly using small, mobile units to employ hit-and-run tactics on security and military targets. These units are usually lightly armed and consistent in avoiding a prolonged direct confrontation with the regime's forces.

Guerrilla warfare is not new to Egypt. What is new are the quality of tactics employed by the Sinaian insurgents, as they fight in a way similar to a combined motorised and/or light infantry and Special Forces.[46] Since early 2014, SP has used a combination of medium (82mm) and light (60mm) mortar artillery, guided and unguided surface-to-surface missiles, guided surface-to-air missiles, heavy machine guns and snipers to cover the advance or the retreat of squad-sized or platoon-sized formations (depending on the operation). SP controlled and/or denied-access to several villages and stretches of land south of the towns of Sheikh Zuweid and Rafah, as well as al-Arish city.

In January 2014, ABM shot down a Russian-made Mi-17 helicopter that belonged to the Second Field Army, killing all five of its crew members. The weapon used was an infrared-homing, surface-to-air missile from the Russian-made *Igla* family (either an SA-16[47] or an SA-18,[48] the newer and more accurate/longer range model). This was the first time in Egypt's history that an armed nonstate actor shot down a military helicopter of the state's army with a missile. In October 2014, SP seized a large number of weapons after a twin attack on military checkpoints in Karm al-Qawadis (Sheikh Zuweid) and al-Arish city. During fighting, SP killed more than thirty soldiers and destroyed an American-made M-60 Patton Tank and an armoured personnel carrier. In November, the organisation issued a propaganda video entitled 'The Charge of the Supporters', in which it showed some of the weapons looted from the regular forces, including medium mortars (82 mm).

The intensity and the scale of the attacks expanded in January 2015, when SP simultaneously targeted eleven military and security posts in three cities/towns: al-Arish, Sheikh Zuweid and Rafah. Simultaneous attacks on such a large of number of targets were by themselves unprecedented in Egypt, even by comparison with the period that British Royal forces and Egyptian insurgents clashed in the Canal cities in the forties and fifties of the last century. During the attacks, SP used light mortars (60mm), heavy machine guns (12.7mm) and various types of improvised explosive devices (IEDs), including remote-controlled ones. SP also used unguided anti-tank rockets from older generations, such as the RPG-7. But their anti-tank arsenal evolved later in 2015 to include Russian-made *Kornet* missiles, a sophisticated anti-tank, laser-guided missile with the ability to penetrate 1,100 to 1,200 millimetres of steel armour protector. The targets of the January 2015 attacks were difficult: well-armed and heavily guarded. They included the camp of the 101st Battalion in al-Arish. This was the headquarters of some of the military forces deployed in the northeast and the place where the military police and intelligence would interrogate suspects. Locals referred to it by the term 'Sinai's Guantanamo.'

In May 2015, SP issued 'The Charge of the Supporters – Part Two', a propaganda video well-documenting simultaneous attacks on seven military targets in April of the same year. But SP's most complex attack came on 1 July 2015.[49] It targeted 21 military and security posts almost simultaneously and succeeded in completely destroying about half of them.

Overall, SP and its predecessor, ABM, conducted over 880 attacks between the beginning of 2014 and the end of 2016, with the most sophisticated attacks, in terms of the quality of military tactics and quantity of the insurgents, occurring in late 2014 and 2015. The majority of the attacks targeted either military or security forces and took place near the northern coastal road between al-Arish and Rafah. Other attacks were conducted against Israel or soft civilian targets, such as gas pipelines. The organisation was also able to carry out significant attacks outside of Sinai, most notably in Cairo, central Delta, north of upper Egypt and the Western Desert, more than 600 miles away from North Sinai. By the end of 2016, the number of military officers and soldiers killed in North Sinai was estimated to be well-over 1,000 compared with 401 in a five-year low-level insurgency throughout Egypt (1992–6).[50] The number of dead insurgents is even more difficult to ascertain. Adding the numbers provided by the military spokesperson since 2011, dead insurgents will exceed 3,000 (compared with 425 in the 1990s' insurgency). However, the identity of the deceased persons is contested. Some of them appear to have been civilians killed in aerial or artillery bombardments, and detainees held by the security forces before their names were listed as 'takfirists'[51] killed during combat. This is in addition to revenge killings by the regime's forces, which usually target the relatives of known insurgents, following a successful attack.

In terms of their resources, arms, training and recruitment are the three most important pillars bolstering SP's military activities. The ruling regime in Egypt accuses Libya and Gaza of being the source of arms, and Turkey, Qatar, Israel and the United States of conspiring with the insurgents.[52] In addition to the intermittent flow of arms from Libya earlier in 2012, many of the SP's arms are taken from the regular forces during local attacks. SP draws its military capacity from three categories within its membership. Defected members of the Egyptian army and security forces belong to one category. It included defectors ranking from conscripted privates to commissioned lieutenant-colonels who served in the Special Forces, Air-Defense, the Navy, Central Security Forces and the Police.[53] The second category is battle-hardened insurgents trained in foreign combat zones, including Gaza, Syria and Iraq. 'We came to you [al-Sisi] from the Levant', said one of the insurgents in the propaganda video entitled 'The Charge of the Supporters – Part Two'. A third category is persistent local insurgents, who accumulated significant experience over the past decade of both combating regular forces and building logistical support networks. Given the counterinsurgency blunders and the political environment, SP was able to recruit from both the Sinai Peninsula and the Nile Valley, albeit in small numbers. 'In my village, we knew of five young men who joined the *Ansar* [ABM] in 2012. Now [in 2015], we think they are over 30', said a local journalist.[54] In addition to direct recruitment, SP has an effective spy network. 'They [SP insurgents] are warned before the Apaches or the tanks reach their targets . . . they have too many friends here', said another local.[55] The sense of grievance, bitterness and even animosity, towards the regime's forces, primarily a product of counterinsurgency policies, is one of the most important factors in creating these 'friendships'.

Mediocre counterinsurgents

Cairo's counterinsurgency policy in Sinai was built on three pillars: repression, informants/intelligence and propaganda (RIP). The RIP policy has varied in its intensity and duration over the last decade. Intensive, reactive and, mostly, indiscriminate repression was the hallmark of the policy in the troubled north. The intensity and indiscriminatory features had a rational, yet also immoral/illegal, purpose: to terrorise and hence subdue a population perceived as potentially rebellious. However, in many cases, especially when the repression was reactive, the intensity and the indiscrimination coalesced into irrational acts of revenge, collective punishment and superiority complexes.[56] Tactics used included torture of suspects, extra-judicial killing of suspects and detainees, demolition and burning of homes, forced evacuations, destruction of property and farms, and the use of heavy artillery and aerial bombardment in residential areas.[57]

As a result, the official numbers of allegedly dead insurgents exceeded the high-end estimates of the size of the insurgency. A military spokesperson claimed to have killed well-over 3,000 'takfirists' since 2011.[58] They included an alleged 438 in ten days in September 2015, 232 in four days in August 2015, 241 in four days in July 2015 and 'more than 170' in a week in February 2015.[59] Sinai's Human Rights Committee in the Egyptian Observatory for Rights and Freedoms (EORF) and the Sinaian Observatory (SO), two local NGOs critical of Cairo's human rights record in Sinai, have a different opinion. Both report that some of the dead suspects were in the custody of the military or the security forces. A well-publicised case is that of Ammar Youssef al-Zer'ai.[60] The military spokesperson declared that he was detained with other 'takfirists' on 8 November 2014. He was then declared killed in combat during a military operation on 14 November 2014. Another infamous incident occurred in January 2017. Ten teenagers were killed, after an SP attack on two security checkpoints in al-Arish city.[61] The military spokesperson declared that they were responsible for the attack and were killed in combat. However, it was shown later by their lawyers, relatives and friends that they were detained as early as October 2016 and were never released.[62] This was followed by demonstrations and calls for a general strike in al-Arish. All of the deceased teenagers belonged to clans and neighbourhoods from which some of the SP commanders hail.

Overall, an EORF report claimed that the military has extrajudicially killed 1,347, detained 11,906 and forcibly deported 26,992 persons between September 2013 and June 2015.[63] Human Rights Watch reported the large-scale destruction of at least 3,255 buildings, and that 'the Egyptian army provided no written warning of the impending evictions and that many residents heard about the coming demolitions from army patrols, neighbours or media outlets'.[64]

Other components of repression had a major impact on the local population. Aerial bombardment of small villages and residential towns by F-16 fighter jets and Apache helicopters has led to loss of civilian lives, including children. More problematically, 'collateral damage' is highly tolerated within the culture of Egypt's armed institutions (military, security/police and intelligence). Hence, denial and further repression are the most common responses to the local journalists who expose the damage. Perhaps the most famous case is that of Ahmed Abu Deraa, an award-winning, pro-military journalist who just gave a one-off different story from that of the military. In September 2013, the military spokesperson declared that they bombed al-Maqat'a village. According to the spokesperson, ten insurgents were killed, and an arms warehouse was destroyed. Abu Deraa, who descends from the village, reported that there were no dead insurgents, five civilian houses and the village's mosque were destroyed, and that a widow and three children were injured. Following that report, Abu Deraa was arrested, declared a 'terrorist' in the media, and eventually faced a military tribunal that sentenced him for six months for 'disseminating misinformation'.

Propaganda is another pillar of Cairo's counterinsurgency strategy. But the quality of the propaganda has had a negative impact on the regular military's credibility. For example, Kamal Allam, an

SP commander who was several times declared dead by the regime, keeps on showing up in new SP propaganda videos. The same applies to Shadi al-Menei, a commander who was also declared dead several times, before SP published a picture showing al-Menei reading his own obituary on a laptop.[65] The military's inaccurate and misleading information applies to statements on operational developments, deaths of soldiers, civilians and insurgents, collateral damage and even inducements for cooperation. The communication policies of the military are based on a 1960s' assumption that the ruling regime has a monopoly on information and narratives. That assumption is not only shattered by twenty-first century social media and a handful of local journalists and activists, but also by well-documented videos and images produced and disseminated by the insurgents themselves. One of many examples is that of the Egyptian navy frigate targeted by a Kornet ATGM in July 2015. The official narrative claimed that the incident happened when a coastguard motorboat was chasing terrorists, and then an 'accidental fire' erupted in the motorboat. SP released a video to show the details of the operation. It involved one insurgent targeting the frigate from a hill with a Kornet ATGM. There was no chase, no motorboats and no 'accidental fire'.

Moreover, the pillars of RIP are in many ways undermining each other. The type of repression used, with the credibility crisis emerging from mediocre propaganda, undermined cooperation with the regime's forces, as well as local intelligence. 'When you go there [to a military checkpoint] to tell them that there is a bomb beside your house, they tell "you planted it you dirty Bedouin" and then beat or arrest you', said a local resident of Sheikh Zuweid recounting the experience of his neighbour.[66] But in addition to the contradictions of RIP, the quality of the soldiers' training and morale is also problematic. Many of the soldiers serving in North Sinai are conscripts sent to the region in the last few months of their service to ensure limited desertions. This type of soldier is not a match for ideologically committed, locally-rooted insurgents. The regime and its media accuse the insurgents of being foreign mercenaries. But local perceptions in Sinai are far from that. The 'mercenary' versus 'martyr' duality, put forward by the official propaganda to discredit the dead insurgents and glorify the dead soldiers, resonates well in Cairo but not necessarily in Sinai. And in many Sinaian towns and villages, some of the locals perceive the duality in a different way: the 'mercenary' is the soldier coming from the Nile Valley to demolish their homes and loot their property and the 'martyr' is the local Bedouin fighting and dying in the ancestral land he could not legally own due to Cairo's enforced policies in the Peninsula.[67]

Currently, it is almost impossible to survey the Sinaian population and freely gauge the levels of popular support. But each side attempts to magnify the indicators of support in their propaganda. In August 2013, a video released by ABM (and confirmed by local journalists) showed a massive funeral, with a motorcade of hundreds of pick-up trucks and four-by-four SUVs, to bury four members of the organisation who had been killed by an Israeli drone while trying to launch a missile attack against Israel. In September 2015, another convoy of at least 20 pick-up trucks carrying armed men passed through villages, while raising the flags of SP. The villagers (including children) were greeting the insurgents on both sides of the road. This does not mean that the ideological extremism or the brutal tactics of SP are popular. Tribal- and grievance-based explanations may provide better understandings. But what it does mean is that the 'contest for the loyalty' is not being won by the incumbents.[68] And, more problematically, that there are no indicators for a policy change in the near future.

Hospitable environment

On a national level, the unruly escalation of Egypt's political crisis since the July 2013 military coup, and the unprecedented repression levels that followed, had major implications for the security situation in Egypt, in general, and Sinai in particular. The bloodiness of the Rabaa massacre (about 1,000 fatalities)[69] and other mass-killings such as the Presidential Guard on 8 July 2013 (51 fatalities), Nasr city on 27 July 2013 (106 fatalities) and al-Nahda Square on 14 August 2013 (87 fatalities), are likely to impact on Egypt's present and future, just as politically motivated massacres did elsewhere.[70] Although Sinai's crises are much older than Cairo's 2013 crisis, the latter had negative consequences for the former. First, Cairo's crisis and its outcomes led to the rise of eradicationist factions and policies within the security and military institutions.[71] These factions believed that Sinai's crisis – among others – could be resolved by eradicating opponents and subduing dissents via almost exclusively brutal force. The leaders of such factions believe that the main mistake of Mubarak and Field-Marshal

Tantawy is that they were 'lenient'.[72] Second, the post-July 2013 repression campaign created an environment wherein arms became an essential tool for both survival and perceived 'justice'. In the words of Yehia Akeel, a former MP representing North Sinai in the since-dissolved Consultative Council (Upper-House):

> Many of the young men carrying arms in Sinai today are not affiliated with any organisations. For them, the coup and what it unfolded meant that they are back to the pre-2011 days of torture and imprisonment without charge. And they prefer the desert and the gun to that.[73]

Third, SP insurgents significantly capitalised on such an environment. As noted above, the organisation is highly sensitive to local developments. Hence, it changed its narrative to bolster recruitment efforts, its pool of supporters, the legitimacy of its existence and the necessity of its brutal tactics.

Battlefield performance: Operation Sheikh Zuweid

On 1 July 2015, SP executed the most complex military operation conducted by an Egyptian armed nonstate actor in the last one hundred years. Throughout that day, 21 guerrilla formations – each ranging between 7 and 150 fighters – attacked 21 military and security targets in the three northeastern coastal districts of Sinai: al-Arish, Sheikh Zuweid and Rafah. The guerrillas started by attacking the first 15 targets almost simultaneously at 06:30. Al-Arish Officers Club, al-Rafai and Abu Sedra military checkpoints (both in Sheikh Zuweid) were hit by SVBIEDs, while the rest of the 12 targets[74] were attacked by a mix of light and medium mortars, ATGMs, unguided RPGs, heavy machine guns and assault rifles. At about 10:00, SP guerrillas took control of about ten of the 15 targets, partly encircling the town of Sheikh Zuweid from the south, east and southwest. While SP formations were still engaging the regular forces north of Sheikh Zuweid town and inside it, SP shelled five additional military checkpoints in Sheikh Zuweid administrative district and detonated IEDs in two other checkpoints. By late morning, there were more than 150 insurgents inside the town taking up positions in the streets and alleyways leading to the heavily fortified Sheikh Zuweid Police Station with the aim of storming it. Abu Khalid, a resident of Sheikh Zuweid who eye-witnessed the attack, said:

They were very organised. The ones in 4x4 vehicles were field commanders giving orders. They had red epaulettes and insignias, walkie-talkies and no masks on . . . The low ranks were teenagers. And they were mainly planting IEDs around the blocks . . . Snipers took positions over the rooftops of the buildings . . . they were in control of the town and it was scary.[75]

The insurgents were also able to station at least five anti-aircraft autocannons on pick-up trucks on the southern and western edges of the town, and one in the centre not too far from the police station.

By late afternoon, the aerial counterattack by the regular forces had started. Initially, the United Command of the East of the Canal[76] used primarily Apache helicopters and ground troops to target its lost positions. This proved problematic for two main reasons. SP rapidly dug in. It planted IEDs on potential attack routes, extensively fired back and its anti-aircraft fire made the mission of the Apaches too difficult. UCAVs[77] and F-16 fighter jets were then used extensively to target the lost positions, as well as small guerrilla formations inside the town. In reaction, SP started burning hundreds of car- and truck-tyres, creating a black smokescreen over some of their positions. This did not undermine the lethality of some of the airstrikes. Abu Mansour, another resident who witnessed the fighting, said

> We counted four 4x4 destroyed vehicles, two burned trucks carrying anti-aircraft autocannons and 16 bodies. My cousins saw the Jihadists carrying at least six wounded fighters from the al-Hamayda Elementary School . . . [which] they turned into a makeshift field-hospital and they had their own ambulance cars.

'I think by around 18:30, most of the *dawa'ish*[78] had retreated . . . there was less bombing and almost no firing', said Abu Khalid.

It is unclear when exactly the SP commanders took the decision to retreat into their strongholds. But it seems to be around 18:00. At that time, the Apaches returned to the battlefield to chase the retreating convoys towards the east (Rafah) and the south (Central Sinai) of Sheikh Zuweid. However, regular armoured and infantry units could not enter the town of Sheikh Zuweid until late in the evening of the next day: Thursday, 2 July 2015. This was mainly due to IEDs and mines planted on the routes.

How 'Sinai Province' fights

As outlined above, the survival of Sinai Province is primarily, though not exclusively, based on how the organisation fights; having unprecedented combat performance compared with all other Egyptian armed nonstate actors. SP did not diverge far from the iALLTR *modus operandi* employed by IS in building the combat capacities of its satellite 'provinces'. There were some peculiarities, however. SP had limited capabilities to raid large weapon depots of the regime, compared to IS units in Syria and Libya. But how does SP manage to stay in the fight under increasingly unfavourable operational conditions? And what kind of tactics give it an 'edge', compared with other Egyptian insurgents? Also, given its limited resources, what strategy does it employ to attain its objectives? This section attempts to address the previous questions by overviewing the categories of tactics employed by SP, analysing its operational performance in Sheikh Zuweid, and its overall strategy based on its publications and statements.

The tactical edge

SP claimed responsibility for 286 operations in its first year of existence (November 2014 to October 2015),[79] and another 476 operations in its second year of existence (November 2015 to October 2016). This represents about a 66 per cent rise in operational activity (but not necessarily in lethality).[80] In the third and the fourth years of its existence (November 2016 to October 2017 and November 2017 to October 2018), SP claimed responsibility for 319 and 229 operations. The last figure is slightly below its first-year record.[81]

After surveying the details of the operations claimed by SP in Northern Sinai between November 2014 and October 2016, a few tactical patterns emerge. Table 5.1 overviews the categories of tactics used by SP in its first year.[82]

In November 2015, SP issued its first annual review of military operations, following the trend of some of IS 'provinces' and mirroring the practices of regular militaries. SP claimed that from 25 October 2014 to 15 October 2015 (corresponding to the Islamic lunar year of 1436) it had killed more than a thousand pro-regime elements, ranging from army and police officers and soldiers to local tribesmen accused of being 'informants' or 'collaborators'. It also

Table 5.1 SP Reported Categories of Tactics, November 2014 to October 2015[85]

Category of Tactics	Reported Usage
Close-quarter Assassinations and Assassination Attempts[83]	167
IEDs	147
Light and Medium Artillery	81
Unguided Rockets	38
Suicidal Guerrilla Formations (SGFs)[84]	32
Guerrilla Formations	18
Anti-aircraft Autocannons	15
SVBIEDS	12
Snipers	11
ATGMs	7
MANPADs	2
VBIEDs	1
Total	530

claimed to have destroyed more than 140 tanks and armoured vehicles and to have razed 30 military/police headquarters and houses of alleged informants. The metrics were listed under six categories: 'Hunters of Armoured Vehicles', 'Targeting Individuals', 'Incursions and Destroying Headquarters', 'Demolishing Forts', 'Special Operations' and 'Spoils of War.'[86]

At the tactical level, the organisation relied heavily on IEDs (28%) and close-quarter assassination tactics (32%). The IEDs targeted armoured vehicles and stationary checkpoints. It allegedly destroyed 24 American-made M-60 tanks, 17 M-113 armoured vehicles and 17 Hummers.[87] The organisation has claimed that it assassinated 130 civilians accused of being 'informants', via exclusively close-quarter assassination/combat tactics.[88] SP also used Russian-made 'Kornet' anti-tank guided-missiles and unguided RPG-7 rockets to destroy seven tanks and armoured vehicles. But most of the damage done by SP was listed in their first annual report under the category of 'incursions': complex guerrilla operations spearheaded by one or more

SVBIEDs, attacking security and military stationary checkpoints and headquarters. The alleged death toll resulting from these operations was around 800 soldiers.[89] In terms of 'spoils', the organisation was able to capture significant numbers of medium and light weapons from the regular forces. It showed many of these in its propaganda videos, including medium mortars (82 mm), two ZU-23 anti-aircraft autocannons (23 mm), five DShK heavy machine guns (14.5 mm) and over 50 AK assault rifles.[90] Overall, the figures are not far from the reports published by the few remaining local journalists and, in some cases, some of them match the results of attacks officially declared by the military spokesperson up to 2015.[91] However, the number of death tolls reported by the military spokesperson vary significantly from SP.

In SP's second year (November 2015 to November 2016),[92] there was a significant increase in armed operations compared to its first year, despite intensively sustained, army-led counterinsurgency operations. Also, there was a significant change in SP tactics and usage of weapons. Table 5.2 overviews the categories of tactics used by SP in its second year of operation.

Table 5.2 SP Reported Categories of Tactics, November 2015 to October 2016[93]

Category of Tactics	Reported Usage
IEDs	263
Snipers	110
Close-quarter Assasinations/Attempts	74
Guerrilla Formations	38
Unguided Rockets	31
ATGMs	12
SVBIEDs	11
VBIEDs	10
Suicidal Guerrilla Formations (SGFs)	4
Anti-aircraft Autocannons	4
Light and Medium Artillery	3
MANPADs	0
Total	**560**

As Table 5.2 demonstrates, SP has significantly increased its reliance on IEDs, sniping, VBIEDs and attacks by guerrilla formations. Usage of IEDs in reported attacks almost doubled from 146 to 263. The same observation applies to attacks with guerrilla formations (from 18 to 38 attacks). VBIED and sniper attacks significantly increased, each by tenfold. In December 2015, SP issued a report illustrating the 'graduation' of a group of trained snipers. But the pictures in the report showed only three armed, masked men in full combat fatigues. They were training with variants of Russian-made Dragunov sniper rifles, with various silencers and telescopic sights, as well as variants of AK-47 assault rifles. This reflected SP's investment in utilising sniping and it was followed by a sniping campaign, the maximum intensity of which was between January and August 2016.

On the other hand, decreases in the reliance on artillery, MAN-PADS, close-quarter assassinations and SGFs were also observed. Losses of mortars were reported in several statements issued by the regular military, which can partly explain the decreased usage. SP has exhibited a limited number of MANPADS in their propaganda videos and used them (MANPADS) only twice since its establishment. Hence, the decrease is probably due to the very limited arsenal. But the decline in close-quarter assassinations merits attention. Informants are a relatively easy target. Close-quarter assassinations (CQA) are not too complex compared with other tactics. And SP has no shortage of handguns, silencers, knives, assault rifles and other weapons used in CQA. But SP has consistently targeted local informants and collaborators (167 reported attacks in 2014–15), and, to a certain degree, succeeded in undermining the security forces' local intelligence network in its first year. The impact of that was to deter local collaborators and decrease the number of informants. This partly explains the significant decrease in CQA numbers by more than a half (from 167 to 74 attacks).

A final observation on tactics is in regard to the use of unguided rockets. The numbers of attacks slightly decreased from 38 in 2014–15 (first year) to 31 in 2015–16 (second year). But unlike in the first year, the attacks in 2015–16 were concentrated in a specific timeframe. They occurred mainly between December 2015 and February 2016. Then there was a long lull, up to October 2016 when SP attacked Al-'Auja/Nitsana Egyptian–Israeli border crossing with two Grad rockets. It seems that SP had suffered from a short supply of unguided rockets in the summer and autumn of 2016.

In terms of causalities, SP publishes estimated numbers of its victims in its monthly and annual military metrics and reports. In many of these reports, the personal details of the victims are published, especially in the cases of alleged informants and high-ranking officers. In January–February 2016 (corresponding to the Islamic, lunar month of *Rabi' al-Thani*), SP issued its monthly 'harvest of military operations', declaring the alleged destruction of 25 armoured/vehicles (including tanks, minesweepers and bulldozers) and the alleged killing of over 100 soldiers.[94] This was done via an overwhelming reliance on IEDs (59 per cent of its operations), followed by guerrilla formation attacks (20 per cent) and then by snipers (12 per cent). The rest of the fatalities were due to close-quarter assassination of informants (9 per cent).[95] In the annual 2015–16 military metrics (corresponding to the Islamic, lunar year of 1437), SP has claimed to have killed and wounded over 1,000 army and police affiliates, including the assassination of 50 informants and tribal militiamen, and having either damaged or destroyed 246 armoured vehicles, include 29 tanks.

In terms of armour, SP was able to capture a tank and two armoured vehicles in the summer of 2016. SP fighters were shown manoeuvring with the captured tank in a propaganda video released in August 2016 and entitled 'Desert Flames'.[96] In January 2017, SP captured in broad daylight another armoured vehicle near the centre of al-Arish city, killing all four of its crew members. However, there were no reported attacks in which SP used armoured vehicles except briefly. During the abovementioned attack in al-Arish city on 11 January 2017, SP fighters captured an armoured vehicle and used it immediately to attack another armoured vehicle that belongs to the security forces. However, the latter was able to manoeuvre and withdrew. The lack of armoured usage by SP is primarily due to the control of the skies by Israeli and Egyptian air-forces (including with UCAVs).

Battle of Sheikh Zuweid: strengths and weaknesses

On an operational level, the 2015 Sheikh Zuweid battle provided some insights into the strengths and the weaknesses of SP's combat performance, and therefore presented valuable lessons on how it might be countered. Regarding the strengths, the organisation had certainly diversified and honed its military capabilities, compared with any other nonstate actor that previously operated in Egypt. It was capable of the operational planning of 21 almost-simultaneous, diversified attacks and of briefly occupying heavily fortified military positions and headquarters. The organisation also skilfully combined sniping, IEDs, SVBIEDS, ATGMs and anti-aircraft autocannons almost to their maximum effect. Likewise, SP was quite strong in terms of their ISR and local knowledge. All five eyewitnesses interviewed agreed that the insurgents were mainly locals from the southern villages, with a distinct northeastern Sinaian accent. Hence, they knew where to position their snipers, where to plant their IEDs and where to locate local informants.

Regarding the weaknesses, the battle has shown that SP had a major manpower crisis. The organisation did not have enough fighters to secure positions. At some of the targeted checkpoints, SP was outnumbered by at least 20-to-1, while continuing to remain on the attack. SP also suffered from a shortage of trained recruits. Hence, recruitment propaganda and training facilities were crucial for organisational survival. More importantly, the battle shows that SP had significant problems with local support in urban areas. The organisation did not formulate any standard operating procedures to deal with these problems during battles. Testimonies of residents show very different patterns of interaction between SP fighters and urban residents. They ranged from limited cooperation under coercion, political and religious argumentations and attempted/failed persuasions, and different forms of defiance and resistance, some of which developed into fistfights. One resident reported that Muhammad 'Ubada Quwaydar, a farmer, had an argument with two snipers who took up positions on the rooftop of his property. The argument developed into a fistfight and Quwaydar pushed one of the snipers off the roof of the building. The sniper was killed instantly, and the other sniper shot Quwaydar.[97]

Moreover, as in other IS Provinces, SP had a major weakness in dealing with high altitude fighter-jets like the F-16s, as well as with UCAVs. These problems were even more significant when SP controlled a relatively large urban territory (as opposed to a small neighbourhood). Finally, the Sheikh Zuweid battle showed that SP suffered from a crisis in strategic planning. As with the case of ISIS in the 2014 battle of Mosul,[98] the objective of the operation was constantly changing based on battlefield developments. SP was operationally planning to take over the town and, for a few hours, it succeeded in doing so. But there was no strategic planning on how to hold on to the town, while being massively outnumbered and outgunned, and with outdated air defences, limited supply lines and limited local support.

After Sheikh Zuweid: a strategy?

Based on the pattern of operations over the last few years, it seems that SP has undeclared offensive and defensive strategies suitable to its structural weaknesses and limited resources. After the Sheikh Zuweid battle, SP understood that conducting operational-level *modus operandi* like SCCLC to occupy cities and towns is beyond their combat capacities. Hence, their offensive strategy aimed to expand the scale of operations and areas-denial; as opposed to occupation of towns. The defensive strategy aims for endurance and survival. For 'endurance', three operational objectives can be identified: recruit, area-deny and blind. The propaganda videos and audios, narratives, segments of the ideology and the high sensitivity to local grievances, local culture and socio-political developments all aim for recruitment and radicalisation. This is crucial to bolster SP's ranks and alleviate its manpower crisis. The new recruits need training facilities. Hence the attempted control and area-denials of villages, swathes of the desert, networks of tunnels south of Sheikh Zuweid, Rafah and al-Arish is critical for organisational survival. The reliance on foreign-trained individuals is a less feasible and a more costly option for SP, at least since 2016, due to regional security developments.[99] Still, Abu Hajar, the leader of SP, declared that 'emigration' (self-recruitment of foreign fighters) to Sinai is open.[100] The third aim is to keep 'blinding' the regime's forces by targeting their local networks of informants, including some of the informants protected in Cairo.[101] This is done regardless of the 'informants' tribal and clan affiliation, as the list of victims includes figures from the same tribes and clans as the SP commanders.

Regarding offence, SP aims for attrition and fear − in line with other IS Provinces that shifted back to guerrilla and terrorism tactics. The sustained engagement with the regime's forces in a long war of attrition is similar to the strategy employed by the relatively under-resourced IS units that previously operated in places such Diyala in Iraq or Hama in Syria. The 'fear' objective is attained by using mass-casualty terrorism and subversive tactics to lower the morale of the regular soldiers, as in a textbook Maoist insurgency. This is done through the sustained targeting of the incumbent's forces, the brutality of executing captured soldiers and officers to instil fear, with the aim of destroying or undermining the soldiers' will, though not necessarily their capacity, to fight. The offensive strategy is executed via some of the aforenoted urban terrorism and guerrilla tactics (including suicidal guerrilla formations, combinations of IEDs, VBIEDs, SVBIEDs and targeted assassinations). SP has low-to-no conventional capacity. It used combinations of medium and light mortar artillery, guided and unguided surface-to-surface and surface-to-air missiles, heavy machine guns and snipers to cover the advances or retreats of mainly squad- and platoon-sized formations (a company-sized formation was reported only inside the town of Sheikh Zuweid during the July 2015 urban battle). The 'quasi-conventional' tactics are usually employed when the objective is to control territory.

The future of Sinai Province

'The operations in Sinai are over . . . and we could have finished terrorism there in six hours', said General Ahmed Wasfy, the Commander of the Egyptian Second Field-Army on 2 October 2013 in an interview.[102] The 'six hours' became more than seven years. Although the number of claimed SP operations in the third (November 2016 to October 2017) and in the fourth years (November 2017 to October 2018) declined to 319 and 229 respectively; the 'Province' is still capable of launching complex attacks on hard targets in both northeast and northwest Sinai.[103] For example, in January and February 2020, the organisation claimed attacks ranging from Rafah in the far-northeast to Bi'r al-'Abd in the far-northwest of Sinai. The categories of tactics employed included SGFs, ATGMs, unguided rockets, sniping and static/covert IEDs.[104]

The future of SP is linked to the fundamental question of why the insurgency has persisted so far, despite its numerous weaknesses and the relative strengths of its foes. As demonstrated above, SP (and IS in general) defied several established conclusions in the insurgency and counter-insurgency literature. David Galula's conclusion – among other scholars – that if the geography is unhelpful, the insurgency is condemned to failure before its starts, does not fully explain this case, among others.[105] In addition to defying geography-centric explanations, population-centric and state-support explanations also fail to fully explain the persistence of this insurgency. Whereas LAG and other repression-centric explanations are useful in understating the recruitment and radicalisation dynamics at work, and also for comprehending why SP is perceived as a 'lesser evil' than the regime's forces in some districts, they still do not fully capture the earlier expansion and current endurance of the 'Province'. Moreover, the borderline scorched-earth and other unrestrained tactics used by the regular forces since September 2013, and their strategic failure to end the insurgency, present another challenge to conclusions in the literature. Morality and legality aside, several studies have concluded that scorched-earth tactics are one of the most effective strategies for defeating guerrillas.[106] This remains far away from the case of SP.

SP endured primarily due to other reasons. The relatively successful build-up of SP's military capacities from its predecessors in the early 2000s to the current phase was a decisive factor, and one that other Egyptian insurgents failed to achieve. The innovative tactical combinations and strategic shifts between guerrilla and terrorist ways of warfare, as well as honing the planning and execution of these tactics, undermined both the morale and the capacities of the regular forces; a historically incompetent force with limited success in conventional warfare[107] and counter-insurgency campaigns.[108] SP was also able to survive – in part – due to the policies of the regime within the peninsula and outside it. Even as the insurgents waged an unusually effective guerrilla war, the regime waged an unusually ineffective counterinsurgency campaign, consistently marred by gross human rights violations. In such a context, SP and/or like-minded groups may endure longer, and possibly expand further.

Notes

1. Ahmed Abdul Azim, 'Major General Ahmed Wasfi to "El-Watan": The Military Operations in Sinai are Over', [in Arabic] *El- Watan*, 12/10/2013, Accessed on 12/4/2019, at: https://bit.ly/2y8mbip
2. That attacks involved SGFs, snipers and static, covert IEDs.
3. The chapter quotes various audio statements, video statements and social media posts by IS and SP. Most of these links were only briefly available online before they were removed.
4. 'Egypt's Sinai Rocked by Wave of Deadly Attacks', *BBC*, 1/7/2015, accessed 3/7/2015, at: https://bbc.in/2RwNNEL
5. 'Egypt's Sinai', *BBC*.
6. Locals tend to take shelter beside the MFO's north camp to avoid indiscriminate bombardment by the regime's forces. This is one of the reasons

LURES AND ENDURES: EGYPT

for the relatively limited attacks on the MFO. The insurgents attempt to limit local animosities towards them by reducing attacks on the civilians' unofficial 'safe havens.'

7. On 24 November 2017, over 300 worshippers were killed in an SP attack on the al-Rawda mosque, surpassing the death-toll of the 2015 bombing of the *Metrojet* plane. Worldwide, this was the second deadliest terrorist attack of 2017, only surpassed by the Mogadishu bombing in October 2017. For details see: Omar Ashour, 'Was the Mosque Massacre in Egypt Preventable?', *The Washington Post*, 01/12/2017, accessed 07/12/2019: https://www.washingtonpost.com/news/monkey-cage/wp/2017/12/01/was-the-mosque-massacre-in-egypt-preventable/

8. 'Islamic State Ways of Warfare' Database (ISWD-SinaiY1 dataset and ISWD-SinaiY2 dataset). Compilations and figures are based on the statements of the Media Office of Sinai Province (*al-Maktab al-I'lami li Wilayat Sayna'*), 'The Annual Harvest of Military Operations for the Year 1437' [in Arabic], 10/11/2016; The Media Office of Sinai Province (*al-Maktab al-I'lami li Wilayat Sayna'*), 'The Annual Harvest of Military Operations for the Year 1436', [in Arabic] 12/11/2015. Some adjustments to the figures listed in the 'Harvests' were made. The adjustments were made after comparing them with statements from the military spokesperson, the local media, and verifications by local North Sinaian sources.

9. Ibid.

10. Ibid.

11. '*Bayt al-Maqdis* Publishes a Video of a Helicopter Shot Down in the Sinai', *Al Arabiya*, 27/1/2014, Accessed on 12/4/2020, at: https://bit.ly/2V2FTFa

12. 'Sinai Province: Egypt's Most Dangerous Group', *BBC*, 12/5/2016, Accessed om 13/4/202, at: https://bbc.in/3b8stwQ

13. Between 1948 and 2014, the United States has provided Egypt's military with $74.65 billion in aid, with more than half of that coming since 1979. See: Jeremy Sharp, 'Egypt: Background and U.S. Relations', *Congressional Services Report*, 5/6/ 2014, at: https://bit.ly/34x6fCb

14. The terrorism and guerrilla tactics employed by SP still have more commonalities than differences with other IS Provinces.

15. The escalation of the military campaign against Sinaian insurgents began in September 2013. But the wave of successive military campaigns began in August 2011, in response to a series of border attacks against Israel perpetrated by Sinaian insurgents.

16. For example, Khaled al-Menei, one of the Sheikhs of al-Sawarka Tribe from al-Menei Clan, headed a pro-military tribal militia before his assassination in Cairo in October 2015. He comes from the same clan and tribe of the SP commander Shadi al-Menei and others. Also, senior commanders in the security and military forces also belong to the same tribe as the SP commanders. The most notable example is Brigadier-General Muhammad Salmi Abd Rabbu al-Sawarka, who was assassinated in July 2014.

17. The historical relationship with Hamas will be briefly discussed later in this chapter.

18. This estimate only includes the forces engaged in Sinai or stationed in the Canal Zone from the Second and Third Field-Army, the Border Guards and the Rapid Deployment Special Forces. It does not include the rest of the military (both active and reserve). It does include some of the forces under the Ministry of Interior, involved in combat, such as the Central Security Forces (which has over 400,000 officers, petty-officer, soldiers, and conscripts within its units) and the National Security Apparatus (estimated to comprise over 100,000 officers, petty-officer, soldiers and conscripts).

19. See Mona Zamlout, '*Ansar Bayt al-Maqdis*: We are not Affiliated with Either the Muslim Brotherhood nor al-Qaida', [in Arabic] *Aljazeera* 28/5/2014, Accessed on 13/4/2020, at: https://bit.ly/2XucsNM; Former MP Yehia Akeel, Interview by author, 6 April 2015.

20. G. Gray, 'Internal Security in Sinai', *WikiLeaks*, WikiLeaks cable no. 05CAIR01978, 14/3/2005, retrieved 10/8/2015, at: https://bit.ly/3ceLo9q

21. Interview by author, Cairo, August 2012.

22. As discussed in the previous chapters, Zarqawi's group started with the same name (*al-Tawhid wa al-Jihad*) and operated later under five other titles.

23. Former Islamist Detainee, Interview conducted by the author, Cairo, September 2012.

24. Ibid. As elaborated upon in the first chapter, the term *takfir* refers to excommunication of the 'other'.

25. Former Prison Warden of the 'Scorpion' and Liman Tora Prisons, Interview by the author, Cairo, September 2012.

26. See: Omar Ashour, *The De-Radicalization of Jihadists: Transforming Armed Islamist Movements* (New York, London: Routledge, 2009).

27. Ibid.

28. Allam was arrested in a security sweep in early 2003. He was imprisoned without charge or trial for about eight years in the notorious Abu Za'bal Prison (2003 to 2011). He broke away from Abu Za'bal in January 2011 during the January Uprising. He was accused of subsequently attacking al-Arish Police Station in 2011, before allegedly travelling to Libya and Syria in 2012. He was later seen in a military uniform in al-Arish city in 2014. Ziada was released from prison in 2009, after a three-year detention. He was credited for recruiting and centralising several jihadist cells and factions (both in the Sinai and the Nile Valley) under ABM's organisational structure.

29. *Al-Tahwid wa al-Jihad* – *Bayt al-Maqdis*, 'This is Our Creed', [in Arabic], [Site of publication unknown]: [no publisher], 2009.

30. This conclusion is based on several interviews conducted by the author with local journalists, activists, ex-military and security personnel, and Sinai-based Salafi figures between 2012 and 2015. Some interviewees estimated the number of different networks/groups was as high as seven. However, it is more likely that the same network used different organisational titles at different times.

31. Interview conducted by the author via the phone, September 2014.

32. Interview of an anonymous member of ABM in: Mona al-Zamlout, 'Supporters of Jerusalem: al-Sisi is an Apostate and We Are Continuing

to Fight', [in Arabic] *Al Jazeera*, 28/5/2014, accessed on 13/4/2018, at: https://bit.ly/2XvZkYD
33. This was the title of a series propaganda videos issued by ABM to document its operations in October 2013.
34. Amrun Al-Naquib, 'Sources for 24: The Iraqi "Abu Hajar Al Hashemi" is the Master Planner for the Massacre of the Rawda Mosque', [in Arabic] *24*, 20/11/2017, Accessed on 13/4/2020, at: https://bit.ly/3b5mZmJ; Ahmed Jumaa, '10 Pieces of Information about Abu Hajar al-Hashemi, the Master Planner of the Terrorist Attack on the al-Rawda Mosque', [in Arabic] *Youm 7*, 30/11/2017, Accessed on 14/4/2020, at: https://bit.ly/2JX3ReK
35. Interview in *al-Naba'* newsletter, issue 60, Hijri-dated 22 *Rabi' Awal* 1438, p. 8.
36. See the interview with the IS commander in Egypt (not in Sinai Province) in *al-Naba'* newsletter issue no. 99. 'Interview with the Emir of the Caliphate Soldiers in Egypt', *Al-Naba'*, no. 99, Hijri-dated 8 Sha'ban 1438, p. 8; Former Major in the Central Security Forces who served in Sinai, Interview by the author, December 2016. On the ISR capacities of SP, see the comments of Major-General Salama al-Guhari, the former head of the counterterrorism unit in the Egyptian Military Intelligence in Sayed Turki and Wael Mamdouh, 'How did Sheikh Zuweid Escape the Fate of Mosul and Ramadi?' [in Arabic] *Almasry Alyoum*, 8/7/2015, Accessed on 13/4/2018, at: https://bit.ly/2VninBL
37. Amrun Al-Naquib, 'Kamal Allam: From a Popular Singer to the Emir of *al-Tawhid wa al-Jihad*', [in Arabic] *24*, 28/12/2017, Accessed on 13/4/2019, at: https://bit.ly/2Vi4LYg
38. Hassan Abu Hanieh and Muhammad Abu Rumman, *The Islamic State Organization: The Sunni Crisis and the Struggle for Global Jihadism* [in Arabic] (Friedrich Ebert Foundation: Amman, 2014). The author contacted one of the book's authors, who mentioned that the information is based on an interview he conducted with supporters of ISIS.
39. IS had already declared a 'Caliphate' on 29 June 2014.
40. The speaker was probably Kamal Allam, one of the military commanders of *Ansar Bayt al-Maqdis/Wilayat Sayna'* (Supporters of Jerusalem/Sinai Province or ABM/SP) from al-Arish city.
41. The quality of SP propaganda, and its wide dissemination, is critical for recruitment and radicalisation. But its direct impact on SP's combat capacity is limited to the 'moral force' and recruitment/radicalisation/manpower dimensions.
42. Perhaps the clearest case were the six bombings of 1954 (attacking the Groppi restaurant and five other civilian targets). They were perpetrated by Nasser's Military Police loyalists. The bombings were blamed on the Muslim Brotherhood and Communist factions. Nasser's involvement was exposed by Abdul Latif al-Baghdadi, the former Minister of Defence (1953–4) and a Free Officer member, in his memoirs (see: Abdul Latif al-Baghdadi, *Memoirs of Abdul Latif al-Baghdadi: Part One* [in Arabic] (Cairo: Al-Maktab al-Masri, 1977). His testimony was corroborated by Khaled Mohidyyin, a Free Officer member, and former Egyptian ambassador

in Switzerland during Nasser's time and General Mohammed Naguib, Egypt's former president, in their memoirs as well.

43. This includes a testimony of an alleged army officer regarding the assassination of Brigadier-General Adel Ragaai, the commander of the Ninth Division, which oversees securing the strategic 'Central Zone' in October 2016: 'A Video Clip Ignites a Controversy over the Role of "Ali Mamlouk" in the Assassination of an Egyptian Officer', [in Arabic] *Orient Net*, 30/10/2016, Accessed on 13/4/2018, at: https://bit.ly/3chrRp8

44. Omar Ashour, *The De-Radicalization of Jihadists: Transforming Armed Islamist Movements* (New York, London: Routledge), 2009, pp. 45–63.

45. Omar Ashour, 'Collusion to Collision: Islamist-Military Relations in Egypt', *Brookings Paper* no. 14 (March 2015), pp. 9–16.

46. Some elements of ABM/SP were defectors from the regular army, most notably Special Forces Major Hesham al-Ashmawy, Special Forces Captain Emad Abd al-Halim, Naval Captain Ahmed Amer (who looted the patrol combat-boat '6th October' from the Naval Base in Damietta in November 2014 and then declared on the radio that the boat belongs to the 'army of the Islamic State') and others.

47. US Department of Defence's designation for a 9K310 Igla-1E missile.

48. US Department of Defence's designation for a 9K38 Igla missile.

49. The Battle of Sheikh Zuweid is analysed in detailed below.

50. Ali Abdul Fattah, *The Religious Situation in Egypt* [in Arabic] (Cairo: al-Ahram Center for Strategic Studies), 1996, p. 71. The numbers are based on compilations found in open source publications, official statements, insurgents' statements and the author's own database on Sinai's insurgency gathered between 2010 and 2015. The figures are estimates, given the difficulty of verifying them. The credibility of the number of deaths given by the military spokesperson and the SP insurgents is highly questionable. At the time of writing, an independent investigation by researchers, journalists, local and international NGOs is almost impossible due to the conditions on the ground.

51. This term is used by the regime to describe the insurgents. It does not correspond to the common scholarly understandings of 'Takfirism' as an ideology. It is more used by the incumbent's forces/media as a derogatory term to delegitimise the insurgents.

52. See for example, the accusations of Major-General Rafiq Habib for the United States on al-Jazeera programme aired on 20 August 2015: 'Evening Window: an Explosion of State Security and Losses on the Egyptian Stock Exchange', [in Arabic] *Al Jazeera News*, 20/8/2015, Accessed on 13/4/2018, at: https://bit.ly/3cpJ0gr

53. See for example Ahmed Hassan, 'One of Us: The Militant Egypt's Army Fears Most', *Reuters*, 16/10/2015, Accessed on 13/4/2018, at: https://reut.rs/2K6pXLD

54. Local Sinaian Journalist, Interview by the author via the phone, August 2015.

55. Interview conducted by the author via the phone. August 2015.

56. Former Major in Central Security Forces who served in Sinai. Interview conducted by the author. August 2016.
57. 'Look for Another Homeland', *Human Rights Watch*, 22/9/2015, accessed on 24/9/2015, at: https://bit.ly/3b5Tp0q; Former Major-General in the Egyptian Armed Forces (Retired), Interview by author, London, 29 September 2015.
58. The author calculated the number based on the statements issued from August 2011 (Operation Eagle 1) to September 2015 (Operation Martyr's Right 2). The exact number is 3,241. However, many of the statements were removed from the military spokespersons' webpage, suggesting that the figures could be inaccurate.
59. Mohamed Tantawi, 'The Harvest of 11 Days', [in Arabic] *Al-Youm al-Sabi'*, 18/9/2015; 'Deaths for the Egyptian Army', [in Arabic] *Al-Jazeera*, 12/9/2015, accessed on 24/9/2015 at: http://goo.gl/mbw53l; 'The Military Spokesperson: Death of 170 Terrorists in Sinai', [in Arabic] *MBC*, 1/3/2015, Accessed on 24/9/2015 at: http://goo.gl/7NoqC9; A. Atef, 'The Military Spokesperson: 241 from the Armed Elements were Killed in Sinai in Five Days', [in Arabic] *Akhbarak*, 6/7/2015, accessed on 1/9/2015 at: http://goo.gl/VLenZu
60. Eid al-Marzouki, Sinaian Bedouin activist, Interview by author, Doha, October 2015. See also: Eid al-Marzouki, 'The Tragedy of Sinai in the Story of Youssef Zera'i' [in Arabic], *al-Araby al-Jadid*, 6/3/2015, p. 6.
61. 'For the Second Day, Funerals of Youth Killed by the Security Turns into Demonstrations', [in Arabic] *Al-Araby Al-Jadid*, 16/1/2017, p. 1.
62. Ibid.
63. Egyptian Observatory of Rights and Freedoms (EORF), 'Sinai . . . Two Years of Crimes', [in Arabic] *Klmty*, 15/6/2015. http://goo.gl/joky2F
64. 'Look for Another Homeland', p. 13.
65. See picture published here: https://tinylink.net/hBOaY
66. Interview by author via phone, August 2015.
67. Land ownership for Bedouins is prohibited by law in many areas in the Peninsula, especially in Northeast Sinai. The military establishment remains the primary owner and regulator of lander ownership in the Sinai. See for example: Abdul Fattah Barayez, 'This Land is Their Land: Egypt's Military', *Jadaliyya*, 25/1/2016; Yezid Sayigh, 'Above the State: The Officers Republic of Egypt', *Carnegie Papers* (August 2012), pp. 16–17.
68. See for example: David Petraeus, James F. Amos and John A. Nagl, *The U.S. Army/Marine Corps Counterinsurgency Field Manual*, p. 312.
69. Human Rights Watch, 'All According to Plan: The Raba'a Massacre', *Human Rights Watch Reports*, 12/8/2014, accessed on 14/8/2014, at: https://bit.ly/2yWS8KW
70. Globally, massacres with that level of bloodshed usually have major social, political and security consequences. The tragic deaths of 14 protestors on 'Bloody Sunday' in 1972 Northern Ireland engendered 'The Troubles' years and a prolonged military and political confrontation between the

British authorities and unionist paramilitaries, and the Provisional Irish Republican Army and other Republicans militias. The Fatima Massacre, the worst in Argentina's 'dirty war', yielded 30 tragic fatalities in 1976 and led to a prolonged armed and unarmed confrontation between the junta and various opposition groups. In the Arab-majority world, the numbers of Rabaa victims were significantly higher than other massacres such as Liman Tora in Egypt (1957) and Dujail in Iraq (1982), and only comparable to Abu Selim in Libya (1996).

71. Youssef Chaitani, Omar Ashour and Vito Intini. 'An Overview of the Arab Security Sector amidst Political Transitions: Reflections on Legacies, Functions and Perceptions', *United Nations' Economic and Social Commission for West Africa (UN-ESCWA)*, New York: United Nations Publications (July 2013); Omar Ashour, 'Collusion to Collision: Islamist-Military Relations in Egypt', *Brookings Papers*, no. 14 (March 2015), pp. 9–16.

72. Major-General in the Egyptian Ministry of Interior, Interview by the author, Cairo, 13 April 2013.

73. Former MP Yehia Akeel, Interview by author, 6 April 2015.

74. The targets were al-Masoora, Sadoot, Al-Wafaq and Abu-Tawila military checkpoints in West Rafah; Wali Lafi in South Rafah; El-Daraib, Kharuba, Obayydat, Qabr Amir, Isaaf, Al-Bawaba and Shalak military checkpoints in Sheikh Zuweid.

75. Ahmed Bilal, 'New Surprises in the Sheikh Zuweid Battle' [in Arabic] *Almasry Alyoum*, 14/7/2015, Accessed on 13/4/2018, at: https://bit.ly/3cbtYKU

76. A new command structure established by Presidential Decree in February 2015, uniting the leaderships of the Second and Third Field Armies. It was led then by Lieutenant-General Osama Askar, the former commander of the Third Field Army.

77. Most residents interviewed did not know to which country the drones belonged. Some of the residents asserted that they were Israeli drones, primarily due to their level of accuracy and their point of origin (east of the borders). Israeli officials did not confirm or deny their involvement in this operation.

78. Derogatory term used to describe the members of IS. It originates from *Daesh*, the equivalent of ISIS acronym in Arabic.

79. Islamic State Ways of Warfare database (ISWD-Sinai Y1 and ISWD-Sinai Y1).

80. Ibid.

81. 'Islamic State Ways of Warfare' Database (ISWD-Sinai Y4). Based on the SP operations claimed in *al-Naba'* newsletter issues from 103 to 152, which cover this period.

82. 'Islamic State Ways of Warfare' Database (ISWD-SinaiY1).

83. SP's close-quarter attackers (aiming primarily for assassinations) commonly use knives and daggers, various kinds of pistols (with and without silencers), machine pistols, submachine guns and assault rifles.

84. The term used to describe the SGFs in Jihadist literature is '*Inghimassiyun*' (literally the 'plungers' or the 'dippers').

85. 'Islamic State Ways of Warfare' Database (ISWD-Sinai Y1 dataset); Media Office of Sinai Province (*al-Maktab al-I'lami li Wilayat Sayna'*), 'The Annual Harvest of Military Operations for the Year 1436' [in Arabic], 12 November 2015. Some adjustments to the numbers listed in the 'Harvest' were made. The adjustments were made after comparing them with statements of the military spokesperson, the local media and verifications by local North Sinaian sources.
86. Media Office of Sinai Province, 'The Annual Harvest of Military Operations for the Year 1436', [in Arabic], 12 November 2015.
87. Ibid.
88. Using various kinds of pistols (with and without silencers), machine pistols, submachine guns and assault rifles and usually escaping on motorcycles or small cars.
89. Media Office of Sinai Province, 'The Annual Harvest'. See note 13.
90. Ibid. All of the above-mentioned weapons were shown in SP's propaganda videos and photo-reports. In a few cases, this included the serial number of the weapon to show that it was captured from the regular army.
91. These reports have been collected since August 2011 by the author. The death tolls reported by the military spokesperson vary significantly from SP.
92. 'Islamic State Ways of Warfare' Database (ISWD-Sinai Y2 dataset).
93. 'Islamic State Ways of Warfare' Database (ISWD-Sinai Y2); Media Office of Sinai Province, 'The Annual Harvest of Military Operations for the Year 1437' [in Arabic], 12/11/2015. Some adjustments to the numbers listed in the 'Harvest' were made. The adjustments were made after comparing them with statements from the military spokesperson, the local media, and verifications by local North Sinaian sources (see the methodology section in Introduction for more elaboration).
94. The regular military statement only acknowledged the death of 27 officers and soldiers in January 2016.
95. Some of the names of the assassinated individuals were published in the SP propaganda. See: Media Office of Sinai Province, 'The Harvest of Military Operations for the Month of *Rabi' Thani* 1437' [in Arabic], 12/2/ 2016.
96. Mahmoud Al Anani, 'Desert Flames: Sinai Province Confirms "We Are Here"', [in Arabic] *Noon Post*, 3/8/2016, Accessed on 13/4/2020, at: https://bit.ly/3a8lUsW
97. Sayed Turki and Wael Mamdouh, 'How did "Sheikh Zuweid" Escape the Fate of "Mosul and Ramadi"? (News analysis)', [in Arabic] *Almasry Alyoum*, 8/7/2015, Accessed on 13/4/2018, at: https://bit.ly/2VninBL
98. See Chapter 1 for details of the two battles of Mosul (June 2014) and (December 2016).
99. A number of SP fighters are from the Gaza strip. But Gazans in northeastern Sinai are considered to be more local than individuals from the Nile Valley due to tribal, familial and cultural ties. The current SP leadership is claimed as foreign (Iraqi). The claim is made by Egyptian security officials and their affiliated media networks.

100. Interview in *al-Naba'* newsletter, issue no. 60, Hijri-dated 22 *Rabi' Awal* 1438, p. 8.

101. A case in point is that of Khaled al-Menei, who was assassinated in Cairo in September 2015. He is one of the well-known pro-military figures from al-Sawarka tribe and a relative of SP commander Shadi al-Menei.

102. Ahmed Abdul Azim, 'Major-General Ahmed Wasfi to "El-Watan": The Military Operations in Sinai are Over', [in Arabic] *El- Watan News*, 12/10/2013, Accessed on 12/4/2019, at: https://bit.ly/2y8mbip

103. See for example the two reports published on 31 January 2020 and 14 February 2020 by *al-Naba'* newsletter issues no. 219 and no. 221 respectively. See also the report published on 7 May 2020 by *al-Naba'* newsletter issue no. 233.

104. *Al-Naba'* newsletter, issue no. 223, Hijri-dated 3 Rajab 1441, p. 6; *Al-Naba'* newsletter, issue no. 221, Hijri-dated 19 Gumada al-Akhr 1441, p. 5.

105. David Galula, *Counterinsurgency Warfare: Theory and Practice* (London: PSI, 2006), p. 23.

106. Arreguin-Toft, "How the Weak Win Wars: A Theory of Asymmetric Conflict', pp. 94–100; Ivan Arreguin-Toft, *How the Weak Win Wars?* (Cambridge: Cambridge University Press, 2005); Douglas Porch, *Counterinsurgency: Exposing the Myths of the New Ways of War* (Cambridge: Cambridge University Press, 2013).

107. See for example: Kenneth Pollack, *Arabs At War* (Lincoln: Nebraska University Press, 2002), pp. 55–73; Norvell De Atkine, 'Why Arabs Lose Wars?', *Middle East Quarterly* (December 1999), pp. 1–5; Omar Ashour, 'A Not-So-Inevitable Defeat: Combat Performance on the Egyptian Front', [in Arabic] in: Ahmed Hussein (ed.), *The June 1967 War: Trajectories and Ramifications* (Doha: the Arab Centre for Research and Policy Studies, 2020).

108. See for example: Dana Schmidt, *Yemen: The Unknown War* (New York: Holt, Rinehart and Winston, 1968).

6

Agency with a Tactical Edge

Over 100 per cent of the caliphate, I took over quickly. Nobody else . . . I took it over . . . We defeated and took over 100 per cent of the ISIS caliphate.

US President Donald J. Trump, 7 October 2019[1]

Our forces are engaged in daily battles against Da'ish [ISIS].

Adel Abdul Mahdi, Acting Prime Minister of Iraq, 19 April 2020[2]

1,557 attacks in 19 countries; 537 in Iraq, 504 in Syria, 116 in Egypt.

IS 'Soldiers' Harvest', 31 August 2019 to 25 February 2020, issued on 5 March 2020[3]

Horizontal escalation: IS after territorial defeats

By 2020, IS had lost almost all the territories it held between 2014 and 2019. The July 2016 proclamation of 35 IS 'provinces' was now recent history;[4] as were the regiment-sized military formations such as the 'Caliphate Army' and the 'Dabiq Army'. IS had also lost its leader and founder, Abu Bakr al-Baghdadi, and over 40 of its most senior military commanders, in addition to tens of thousands of mid-ranking commanders and low-ranking fighters. Despite that, the organisation endured. Less than a week after its decapitation, the IS Consultative Council chose Abu Ibrahim al-Hashimi al-Quraishi (Ameer al-Mawla) as al-Baghdadi's successor, with IS media outlets announcing him as the 'Caliph' on 31 October 2019. In 2020, IS were still proclaiming 14 'provinces', composed of a territorial-less mix of insurgent organisations and urban terror cells. IS shifted again to mainly guerrilla and terrorism tactics, with limited and intermittent usage of conventional tactics in all four countries examined in this book and elsewhere. In Iraq, the Security Media Cell of the prime minister's office declared that the army, the PMUs, the CTS and the federal police conducted 1,060 counterinsurgency and counterterrorism operations against IS targets in all Iraqi Governorates (except Kurdistan) between 1 January 2020 and 15 April 2020. The figure reflected the intensity, scale and scope of the insurgency,[5] with IS claiming 139 operations in Iraq in March 2020. In 2020, the organisation still retains a battalion-sized formation operating in parts of the Euphrates valley, and smaller formations operating in ten different 'sectors' of the restructured 'Iraq Province'.[6] In Syria in 2020, the organisation retained enough combat capacities for unconventional combined arms manoeuvres conducted by platoon-sized formations in al-Badiya/Homs desert. These included coordination between units employing technicals and ATGMs, and handlers of modified UCAVs. IS claimed 43 operations in Syria in March 2020, down from 68 operations in February 2020. In Libya, the above-quoted IS' 'Soldiers Harvest' did not claim any operations between 31 August 2019 and 25 February 2020. Earlier, in July 2019, IS Libya Province released a video showing parts of its (platoon-sized) 'Desert Company', operating near Sabha. The video-release revealed twelve 4x4 (Toyota) vehicles and technicals, one of which was mounted by a quadruple-barrelled ZPU-4 anti-aircraft gun, giving a quasi-conventional touch to the force. *Al-Naba'* newsletter reports regularly on the fighting between what it calls the 'apostates' forces' (the GNA and Hefter's LNA). Hence, IS could be lying in wait for both sides to further bleed and weaken each other. In Egypt, Sinai Province has perplexingly endured two years of intense military operations dubbed 'The Comprehensive Operation' (started in February 2018). Mockingly dubbed by IS as 'The Failed Operation', a moniker that rhymes in Arabic,[7] SP still sustains an average of about five operations per week. The range of the operations covers the whole coast of North Sinai from the demolished Rafah to the still-standing Bi'r al-'Abd. Up to the time of writing, SP is still capable of area-denying almost the same stretch of villages and swathes of land it had been denying to the regular forces six years earlier (as of November 2014 and denied by SP's predecessors since February 2011). SP claimed 18 operations in Sinai in March 2020.

Beyond the countries and governorates studies here, IS continued pursuing a strategy of horizontal escalation. It sustained operations, repeated tactics, conducted knowhow-transfers[8] and attempted to repeat the iALLTR *modus operandi* (with mixed outcomes) in states such as the Congo (DRC), Mozambique, Burkina Faso, Nigeria, Niger, Afghanistan and Yemen. Despite the territorial losses, IS still thinks of itself as a 'state'; even issuing Wahhabist-style 'advices' and medical alerts for its members and supporters to deal with the COVID-19 pandemic. Overall, Trump's 2019 declaration that IS is '100 per cent' defeated was unrealistically optimistic.

To answer the research questions, the previous chapters have shown how IS was able to build-up its combat capacities and how it effectively employed them to fight, expand and endure in four countries. The level of tactical ingenuity and strategic adaptation of the organisation's 'provinces' and combat units is remarkable in comparison with other ANSAs. In the Arab region, the combat effectiveness of IS surpasses that of many ASAs, considering its limited resources. Previous chapters have also highlighted how the group of case-studies (represented by a sample of IS Provinces) poses challenges to some of the literature on military, insurgency and terrorism studies; as well as to some of the security policies employed in the region. Based on that, this chapter summarises the research findings on how IS fights and what makes it unique in terms of relative combat effectiveness. Following that, the last two sections present concluding observations for future research.

Research findings

Let us return to the central question of how IS fights and figuring out the puzzle, which accounts for IS combat and military effectiveness, current endurance and earlier expansion; despite relatively limited resources. Based on the previous chapters, the four variables – 'combat effectiveness', 'military effectiveness', 'expansion' and 'endurance' – are different and each merits its own explanation. 'Combat effectiveness' is directly related to the tactical and the operational levels of warfare. 'Military effectiveness' is intertwined with the strategic level. 'Expansion' reflects battle- and war-outcome(s)/military success(es) in IS' offensives and counteroffensives. 'Endurance' reflects the outcomes of IS defences and survival campaigns. Endurance and expansion, as explained earlier, involve more than combat and military effectiveness.[9] The following sections outline what is unique and what is not-so-unique about IS combat performance to answer the research questions.

What is not-so-unique about IS combat performance?

IS has benefited from numerous macro-, meso- and micro-level variables, conditions and features. Most of them, however, have been unexclusive; other ANSAs (and certainly other ASAs) have also accessed and benefited from them. Perhaps the macro-level most clearly demonstrates this. The regional context and the dominant regime-type within it – featuring high-levels of brutality, illegitimate/unconstitutional state-violence, corruption and decades-long manipulation of sacred religious texts, as well as ethnonationalist/ethno-sectarian intensive propaganda, narratives and policies – all mattered for IS combat and military effectiveness as well as for its endurance and expansion. In all of the four countries examined (and beyond), IS and its predecessors learned how to recruit, radicalise, 'extremise' and militarily capitalise on the aforenoted ills. However, other ANSAs and ASAs also operated within the same context and capitalised (or attempted to) on the same ills, directly or indirectly.

On a micro-level, the fighter-type mattered. The *dogmatist* types (such as Abu Omar al-Jazrawi in the 2014 battle of Mosul and Abu Firas al-Emirati in the 2017 battle of Raqqa) were critical in suicide-intensive offensives or in defensively breaking momentums, in the cohesion of units and in raising the morale of fighters. The recruitment from significantly large pools of *oppressed* youth helped IS alleviate its constant manpower crisis and contributed to its local intelligence capabilities (for example in Sinai). The empowerment of marginalised *opportunists*[10] was integral for logistics, maintenance, tactical and operational intelligence and other combat-relevant aspects. The *experienced* commanders and fighters – both local and foreign to the operating theatres,[11] and with both formal[12] and informal[13] military training and experiences – were crucial in tactical, operational and strategic leaderships (such as Hajji Bakr and Abu Omar al-Shishani in Iraq and Syria); in strategic and tactical knowhow transfers (such as Abu Nabil al-Anbari in Libya); in training and in overall morale-boosting (e.g. acts that inflamed the morale of already-overzealous fighters, such as al-Bilawi's self-detonation at the beginning of the 2014 battle of Mosul). These fighter-types are by no means exclusive to IS. From Colombia to the Philippines – including Algeria, Libya, Egypt, Syria, Iraq, Afghanistan and elsewhere – other ANSAs have also recruited dogmatists, opportunists, oppressed and experienced fighters, both local and foreign in different quantities and varying quality levels. These ANSAs had mixed combat performances, but for the most part have not attained the relative success accomplished by IS.[14]

The meso-level is where the uniqueness of IS' combat performance can be situated. Nevertheless, IS has had a few 'shared' and *not-so-unique* features and variables at this level, the most combat-relevant of which are 'moral force' and psychological warfare, unit cohesion, autonomisation, transregional combat-experiences, flexible hierarchies and specialisation-focus.[15] The commitment to, and the focus on, specialisations and missions, goes further than al-Muhajr's March 2008 strategic shift to intensify urban terrorism in Iraq.[16] It was also exhibited by Hajji Bakr's decision to restructure the IS Military Council. Flexible tactical and operational hierarchies were almost always based on need and on an unwritten *'urf* (tradition), as opposed to less-flexible internal regulations or standard operating procedures (SOPs). This had its costs and benefits. The costs were demonstrated in the previously mentioned dispute between Abu Saad al-Hadrami and Abu Luqman in Raqqa in 2013[17] and the less-reported case of Abu al-Faruq al-Tunsi, the Tunisian *wali* (governor) of *al-Khayr* Province (Deir Ezzor) in 2014.[18] These costs were sometimes offset by a less-flexible strategic command. The flexible hierarchy did, however, have the benefits of temporarily alleviating the crises of manpower and expertise/knowhow.

What Carl Von Clausewitz calls 'moral force' – or the group of psychological variables and values such as morale and bravery – has proven high among IS combat units, at least in offensives such as Mosul (2014), Ramadi (2015), Sirte (2015), Sheikh Zuweid (2015). This was also true for counteroffensives such as those witnessed in Raqqa City in 2014 (against Syrian rebels), Raqqa Governorate in 2016 (against the Assad regime's forces) and even in Raqqa City in 2017 (a series of neighbourhood-based counteroffensives against the SDF and the Coalition). The 'moral force', the reputation attached to having it (even if it is asymmetric, or untrue, among most units) and the well-broadcast brutality of IS gave the organisation repeated advantages in psychological warfare. Moral force, however, was not an exclusive feature of IS units by any means. ANSAs and ASAs that fought against IS also exhibited high levels of moral force and were determined to annihilate ISIS/IS. These included organisations as different (and as hostile to each other) as the Sunni *Liwa' al-Tawhid*, the

Salafist Army of Islam, the predominantly Shiite PMUs, the secular, ethno-leftist YPG and the partly Salafi-Jihadist DMSC, whilst, among ASAs, the US-sponsored Iraqi CTS stand out.

Strong unit-cohesion is another feature, shared as well by some of the other ANSAs and ASAs in the four countries. It was repeatedly exhibited by IS combat units. Even when both strategic and operational command-and-control had collapsed in battles like Sirte City (2016) and Raqqa City (2017) or when defeat appeared inevitable, as in Sheikh Zuweid (2015) and Derna (2016), IS tactical-level unit-members *mostly* sustained their commitment to each other and collectively fought back, retreated, negotiated, or attempted to negotiate. Out of the 31 interviewees who directly engaged ISIS/IS in battles, only two contested their strong unit cohesion (both in the 2016 battle of Sirte).

A fifth feature, with a few critical ramifications for combat performance, is autonomisation. Many IS units, as small as squad-sized formations, were self-reliant (sometimes reaching self-sufficiency), even when they were cut off from higher command-and-control structures. This enabled them to persistently pursue tactical and even operational objectives under adverse conditions. This partly explains why the organisation kept on expanding – and endured – while suffering sustained decapitation. High-levels of autonomisation were demonstrated in Derna (2016), in Mosul (2017), in the Hajin Pocket (2019) and even in Homs/al-Badiya desert (as late as 2020). The following quote from the IS military commander of Raqqa in 2017 summarises autonomisation:

> [IS] demonstrated a talent for developing weapons, such as aerial weapons, drones, anti-building weapons, sniper weapons, silencers, explosive devices, and other types of weapons . . . and [IS] also divided the [Raqqa] city into small sections that could function individually in emergency situations and could independently pursue their objectives . . . these sections would each have their own supplies and ammunition and would be independent of one another in that regard. They would also function independently in terms assessing locations and organizing troops.[19]

Autonomisation enabled a series of (intended and unintended) developments relevant to how IS fights and exhibited itself in the four countries examined. They include tactical aggression and initiative (such as that displayed during the Raqqa battles of 2013 and

2014), improvisation (demonstrated during the battles of Derna in 2016 and – superbly – in Hajin and Baghuz in 2019), relatively fast aggregation-disaggregation-reaggregation (exhibited for example in the battles of Mosul and Raqqa in 2017 and earlier in the battle of Mosul in 2014) and flexible operational-level planning and rapid reactions (present in the battles of Mosul of 2014 and Ramadi in 2015). Within the flexible operational-level planning, IS combat units have even changed operational objectives from 'raid' to 'raid, fortify, and occupy' during some of the urban battles discussed. Autonomisation, however, is not a unique feature of IS combat units. Other ANSAs in Iraq, Syria, Libya, Egypt and elsewhere have exhibited similar features. Usually organisations that start as small, secretive and illegal cells and use urban terrorism tactics in an adverse environment exhibit different qualitative levels of autonomisation if/once they become conventional or quasi-conventional forces. These include groups as diverse as the Irgun and the Lehi/Stern Gang (which became part of the Israeli Defence Forces at a later stage), Hizbullah, Hamas, *Hay'at Tahrir al-Sham* (Levant Liberation Organisation or HTS) and others.

A sixth non-unique feature is the combat-multiculturalism and multitude of experiences among IS units. Within different combat formations – especially the platoon- to regiment-sized ones – the organisation recruited a motley mix of combat-relevant experiences and backgrounds (former soldiers and officers, ex-convicts, local criminals and smugglers, experienced transnational insurgents and informally/irregularly trained fighters, engineers, technicians, craftsmen, and others) and combat-cultures (including fighters from Latin America and the Caribbean, Sub-Saharan Africa, the Caucasus, the Balkans, Western Europe, Post-Soviet Central Asia and Southeast Asia). The selection of fighters from different local-geographical sectors also bolstered the diversification of experiences among combat units.[20] However, other ANSAs also exhibited that diversity at different levels. They include groups and nonstate armed coalitions such as the GIA, the HTS, the SDF, the DMSC and others.[21]

All of the above-mentioned features have enhanced both the combat and the military effectiveness of IS and contributed to its current endurance and previous expansion. However, the organisation displayed other unique features that reflected its distinctive fighting style and proved decisive in determining the duration of its endurance and the scale of its expansion.

What is unique about IS combat performance?

The uniqueness of IS combat performance lies in the quality and the quantity of its tactical innovations, as well as in its three strategic/ways-of-warfare shifts. The organisation's combat units have also relatively shown – compared to other ANSAs – sustained, high-level capacity of operational planning and execution. These tactical innovations, strategic shifts and operational artistry have made the difference in IS' combat and military effectiveness, endurance and expansion; even when its combat-relevant resources should have yielded none of that. An overview of these findings is outlined below.

On the operational level

I will start with the operational level, as it is the main modification to the two initial hypotheses of this book.[22] Three *modi operandi* were critical to the combat performance of IS: iALLTR (on a state-level), SCCLC (on a town/city-level) and a cult of (operational) counteroffensive (at a battles-level; usually as a series of interrelated urban and suburban battles).

iALLTR

The savvy local intelligence-gathering and covert/infiltration operations, the sustained collective absorption of organisations/factions and recruitment of individuals, the constant looting of arms from ANSAs and ASAs, the reliance on experienced, battle-hardened leaderships and the rapid knowhow transfers of tactical innovations and strategic shifts were all repeatedly demonstrated by IS in Syria, Libya, Egypt and other states. The sustained, transnational combination/coordination of operations within the iALLTR *modus operandi* was unique in its quality. Leadership and knowhow transfers should be specifically highlighted in comparative perspective. In contrast to the relatively common corrupt hacks/incompetent leaderships in ASAs and ANSAs operating in the four countries examined, IS was led militarily by the likes of Hajji Bakr and al-Bilawi. The *rapid* knowhow-transfers between the four countries, within them and beyond them, stands out as well in comparison with any other ANSA. It even stands out in comparison with some of the Arab ASAs, in which disinformation and denial of tactical and strategic

knowledge within and between units is a common practice embed-
ded in the institutional culture(s).[23]

SCCLC

Outside of Iraq, SCCLC was not as successful as iALLTR except in
Raqqa City in 2013 as demonstrated earlier. Almost everywhere else,
it was either an operational failure (such as in most of the northwestern cities and towns of Syria in 2013) or only partially implemented
(such as in Sirte in 2015). When they were successful, 'soften and
creep' operations – as Mao (1937) explained decades ago – ground
down the will and terrorised the foes before launching a conventional attack. They also fixed anti-IS forces in their positions within
the defensive role of guarding or garrisoning the town or city. This
enabled a freer IS activities and preparations in the outskirts of the
targeted town or city (e.g. Mosul 2014 and Derna 2015). When the
conventional (or quasi-conventional) attack was finally executed by
IS – with or without a coalition – it was almost always launched
under conditions of disadvantage (such as being outnumbered and
outgunned). When SCCLC was a failure, 'soften and creep' operations neither terrorised nor grinded down the foes of IS enough.
They just served as early warnings of the credible threat of IS' brutal
domination and therefore mobilised local armed organisations against
it, in a classic balancing act. SCCLC showed that IS is capable of both
(tactical) *coalition-building* and *fighting as a part of an alliance* when it
pragmatically chooses so or needs to (such as the case of Fallujah in
2014). But as Iraqi and Syrian insurgents would learn, IS is extremely
treacherous. 'Liquidation and consolidation' operations were almost
always executed against its former allies and then justified by some
theological/ideological arguments, narratives and propaganda. Hence,
any coalition it had built or joined was short-lived. When it does not
want or need to build a coalition, IS can still pull off an unexpected
military upset sometimes – even with high costs and low gains – as it
did in Ramadi and Sheikh Zuweid in 2015.

A cult of counteroffensives

In all the battles analysed in this book, from Fallujah in the east
to Sirte in the west, IS launched tenacious and coordinated counteroffensives. It did so in most of the lost districts within Fallujah,

Ramadi, Mosul, Raqqa, Derna, Sirte and Sheikh Zuweid. Many of these counteroffensives failed to recapture the liberated territories. Some did not make any (military) sense according to (common) urban warfare planning. Overall, their outcomes were mixed. IS' counteroffensives succeeded in Raqqa in 2014 against Syrian opposition forces; in Sirte in 2015 against the 166th Battalion; and in Ramadi in 2015 against the Iraqi regular forces and their allies. When the counteroffensives failed, they delayed and raised the costs of the operational victory for the liberating forces, such as in Ramadi in 2016 and in Raqqa and Mosul in 2017.

Regardless of outcomes, a 'cult of counteroffensives'[24] is clearly present among IS combat units. Given the repeatedly demonstrated capabilities of IS commanders to coordinate squad- and platoon-sized formations, to utilise conventional and unconventional combined-arms tactics, to adequately manoeuvre and sometimes outflank their foes during some of these counteroffensives; that 'cult' should not be mistaken as simply a mere product of moral force/zeal. It is more likely a by-product of both strategic planning (of how IS units *should* fight) and tactical training. The comments of the IS military commander of Raqqa (2017) and the observations made by a BM military commander in Sirte (2016) are almost identical in describing how IS launched sustained counteroffensives in urban and suburban districts using swarms, manoeuvres, encirclements and infiltrations via deception and concealment.[25] This 'cult of counteroffensives' stands in a stark contrast to how many other Arab ASAs and ANSAs have traditionally fought. On an operational level, it shows another layer of uniqueness in terms of how IS fights.

On strategic shifts (and failures)

At the strategic level, IS has no *viable* grand strategy. Simply put, the organisation's resources are way too limited to achieve its ultimate objective (a Wahhabist-style 'Caliphate' with political-military dominance over some or all Muslim-majority states, and with the capacity to 'invade' others). Regardless of how *well* it organises, distributes, plans and applies these limited resources; quantity overrides quality here.[26] Additionally, the organisation's hyper-extremist ideology and murderous behaviour combined with its limited resources prevented it from achieving preantepenultimate or even propreantepenultimate strategic objectives such as permanently occupying Sunni-majority areas in parts of central Iraq, northern Syria, coastal Libya, the Sinai and/or elsewhere. To use Field-Marshal Montgomery's phrases for explaining strategy, IS excels in the 'the art of fighting' (tactical levels) taking into account its limited resources and other challenges; but it is mediocre in the 'art of the conduct of war' (strategic levels).[27] However, being highly skilled in the art of fighting – without being able to secure any strategic victory – has major military, security, political, social, economic, humanitarian and geostrategic implications; some of which have been previously outlined.

Despite the aforementioned caveats, IS mastered strategic shifts between conventional, guerrilla and terrorism warfare, in ways and at levels that most ANSAs were unable to achieve. As shown in the previous chapters, mastering these shifts was decisive in the endurance of ISI/ISIS/IS and was crucial to their expansion. It resulted in the organisation's adaptations and effectiveness under adverse operational conditions and unfavourable strategic environments. All three shifts were executed – back and forth – in all four countries examined, including all seven governorates of Nineveh, Anbar, Raqqa, Deir Ezzor, Derna, Sirte and North Sinai, and all the cities/towns in which IS combat performance was analysed except in Sheikh Zuweid[28] (i.e. in Fallujah, Mosul, Ramadi, Raqqa, Derna and Sirte). IS utilised these strategic shifts elsewhere with similar outcomes, as well. These shifts were boosted by knowhow transfers and included innovative tactics, discussed in the following section.

On tactical innovations (and successes)

In a 2006 interview, Abu Mus'ab al-Zarqawi linked a potential 'victory' of AQI's war in Iraq to the tactics used by the organisation.[29] He dismissed static, roadside IEDs and mortar tactics as incapable of deciding the war's outcome. Alternatively, he referred to an operational combinations of suicide bombers and guerrilla formations. It was an early indicator of the forthcoming combat-effectiveness of suicide guerrilla formations (SGFs) or the *inghimassiyun*, as well as a host of other tactical innovations previously analysed.

Overall, IS relied on fifteen categories of tactics in its major battles discussed. These categories provide an overview of *how IS fights*. They are listed in summary below:

Table 6.1 IS Categories of Tactics and Strategies of Warfare

Category of Tactics	Common in Operations and Strategies of
1) IED-intensive Tactics	Terrorism, Guerrilla, **IS-Conventional Warfare**
2) VBIED Tactics	Terrorism, Guerrilla, **IS-Conventional Warfare**
3) SVBIED Tactics	Terrorism, Guerrilla, **IS-Conventional Warfare**
4) SGFs Tactics	Terrorism, **IS-Terrorism**, Guerrilla, **IS-Conventional Warfare**
5) Sniping Tactics	Terrorism, Guerrilla, Conventional Warfare
6) UCAVs/Drones/DBIEDs Tactics	Terrorism, Guerrilla, Conventional Warfare, **IS-Conventional Warfare**
7) Unguided Rockets Tactics	Terrorism, Guerrilla, Conventional Warfare
8) Guerrilla Formations Tactics	Guerrilla, Conventional Warfare
9) Artillery Tactics	Guerrilla, Conventional Warfare
10) ATGMs Tactics	Guerrilla, Conventional Warfare
11) Anti-aircraft Autocannons' Tactics	Guerrilla, Conventional Warfare
12) MANPADs Tactics	Guerrilla, Conventional Warfare
13) Assassinations Tactics	Terrorism, Guerrilla Warfare
14) Armoured Vehicles Tactics	Conventional Warfare
15) Tunnels Tactics	Guerrilla Warfare

The organisation has sustainably innovated in the techniques and the procedures to manufacture, develop, weaponise and execute these categories of tactics and their necessary resources. This is especially the case in the first six categories (IEDs, VBEIDs, SVBIEDs, SGFs, Sniping and DBIEDs). As previously shown, the first three were essentially transformed from common urban terrorism categories of tactics into 'IS-common' categories used in terrorism, guerrilla and conventional warfare. The last three were significantly modified and upgraded to affect battle-outcomes or to maintain a high-level of combat-effectiveness against superior forces. Overall, these combinations were decisive in enhancing IS combat and military effectiveness. They affected battle-outcomes, especially in the 2014 and the 2015 battles when the anti-IS forces in the four countries (and beyond) were still unfamiliar with, and unadapted to, IS battlefield innovations.

In Iraq, Syria and Libya, IS kept on improvising, modifying and upgrading within the weapon-armour-mobility triad under adverse conditions, even when its units were on the brink of a decisive defeat (such as in Derna in 2016 and in the 'tent-city' of al-Baghuz in 2019). IS mounted captured artillery pieces on looted tanks chassis, replaced cannons with anti-aircraft guns when it deemed necessary, up-armoured infantry fighting vehicles, mounted BMP turrets on 4x4 vehicles, upgraded T-55 and T-62 tanks with locally manufactured armour (some were even made of shell-casings), converted flatbed trucks into weapon platforms. Its combat units used commercial drones to deliver IEDs and to guide SVBIEDs, utilised SVBIEDs as precision bombs and guided rocket-barrages, converted 7-tonne large SVBIEDs into cruise-like missiles/human-guided land-torpedoes and executed Marine-like infantry-breaching sequences combined with SVBIEDs. They employed commercial GPS applications on civilian smart-gadgets to enhance the accuracy of their mortar-shots, teleoperated sniper-rifles to offset the snipers from the weapons for protection, outflanked tanks with swarming anti-tank kill techniques and pioneered in the execution of several SGF-tactics in both urban terrorism and conventional battles, among numerous other tactical-level innovations.

Three concluding observations on how IS executes these categories of tactics are worth reiterating. First, the organisation has shown consistent usage of unconventional combined arms tactics, with different levels of quality. This asymmetry was clear on the battlefields; the organisation's combat units showed mastery in Mosul in 2017 but mediocrity in Derna in 2016. 'If you retreat to the left, you die with sniper-fire . . . if [you go] to the right you die with the [simultaneous SVBIED] explosion . . . They [IS units] gave you no choice in counterattacking', said a BM commander who fought IS in Sirte and its outskirts, describing a sample of IS unconventional combined arms tactics.[30] These tactics were effective in surprising anti-IS forces in offensives and counteroffensives, until they became familiar with it and the surprise element vanished. These tactics were also effective in defensive operations, mainly by countering the overwhelming advantages of the liberating forces as the combinations of UCAVs-SVBIEDs-SGFs demonstrated in Mosul and Raqqa in 2017. Given their combat-effectiveness, unconventional combined arms are likely to be used in the future by IS and/or other ANSAs.

The second observation is the 'bottom-up', mission-oriented approach that IS employed when executing selected categories of tactics. IS commanders are more likely to have emphasised the operational objectives of a battle or campaign, rather than the 'up-to-bottom' ordered tactics to achieve them. This may need further investigation to ascertain in other comparative IS battles. In any case, it is neither new nor unique per se; rather it dates back to the nineteenth-century Prussian army's battles against Napoleon.[31] A statement about the nineteenth-century Prussian forces accurately describes how IS fought in the battles of Mosul in 2014, Ramadi and Sirte in 2015 and others:

> the Prussian General Staff, under the elder von Moltke . . . did not expect a plan of operations to survive beyond the first contact with the enemy. They set only the broadest of objectives and emphasised seizing unforeseen opportunities as they arose . . . Strategy was not a lengthy action plan. It was the evolution of a central idea through continually changing circumstances.[32]

'Mission-type tactics' were upgraded, perfected and utilised later by the *Wehrmacht* and other highly combat-effective forces, including the Israeli Defence Forces (IDF).[33] However, what makes them unique to IS is that they are uncommon among Arab ASAs and ANSAs alike, most of which prefer centralised planning and clearer up-to-bottom orders.[34] Overall, like the above-mentioned 'autonomisation', the 'mission-type' approach to executing tactics has also enabled and encouraged improvisation, innovation, tactical aggression, manoeuvring and flexibility in tailoring the repertoire of tactics to the operational objective.

Third, IS mostly relies on the tactics that it relatively *excels* at executing in the battlefield (due to accumulated experiences), as opposed to those it is just *capable* of executing (due to available resources). This was well-demonstrated in the cases of the summer 2014 battles of Raqqa Governorate (such as the 17th Division and the 93rd Brigade battles), when IS relied on SVBIEDs to achieve favourable battle-outcomes despite having tanks and heavy artillery. Again, this contrasts well with Arab ASAs' behaviours, where the focus is on buying and using the newest pieces of equipment without necessarily being able to effectively operate them in combat.

Finally, it is important to stress *relativity* in combat-effectiveness and in the quality and quantity of tactical innovations, operational art and strategic shifts. Combat is an adversarial, competitive endeavour. To use a traditional martial arts analogy: to beat a white-belt, a fighter just needs to be a yellow-belt and not necessarily a blue- or a black-belt. Certainly, IS is neither as skilful as the Brandenburg Division of the *Wehrmacht*, nor as ferocious as the Teishin Shudan Division of the Imperial Japanese Army. Nonetheless, the organisation needed to endure and/or expand in the face of the US-supported Iraqi, Libyan and Egyptian ASAs and Syrian ANSAs; as well as against Russian-backed Syrian ASAs and ANSAs and Iranian-sponsored ANSAs in the four countries and beyond; and with relatively limited resources. Up to the time of writing of this book, it had managed to do just that.

Back to the future: explodes, expands, endures, shifts and reloads

Within the ANSAs' 'combat-league', IS has shown unprecedented levels of ingenuity in tactical innovation, adaptability in strategic shift, and relative effectiveness in combat. It also followed a vicious pattern, repeated five times in the last sixteen years (2004–7; 2007–11; 2011–14; 2014–19; 2019–). This pattern usually starts with IS and its predecessors rising and expanding, then enduring, then being knocked down (but not out), then strategically shifting and tactically innovating before finally reloading the same pattern with operational modifications; sometimes under a new organisational title.

Overall, there are a whole host of aforementioned macro-, meso- and micro-level factors that explain why IS militarily endured and previously expanded in Iraq, Syria, Libya and Egypt and elsewhere. But the exceptional difference in the organisation's combat effectiveness is rooted in its strategic shifts and tactical innovations, as well as from the above-mentioned qualifiers in its operational-level art of warfare. The organisation rapidly adapted to changing environments and situations. When it succeeded militarily in territorial control, it would capitalise to boost its limited resources (such as in Nineveh, Raqqa and Deir Ezzor). When it faltered militarily, it would launch tenacious operational counteroffensives, and then shift its strategies and adjust or innovate its tactics accordingly. This is how IS fights. And understanding how it fights is critical for its ultimate defeat. At the time of writing, this has yet to be achieved, and could be a subject of a future research project reassessing and analysing *how to fight IS* and like-minded organisations.

Finally, agency and the type of the agent matter. Regardless of the structural challenges imposed and their impact on limiting resources, agents (both ASAs and ANSAs) managed to enhance their combat effectiveness with whatever limited resources they have had. The pre-1967 IDF and the post-1979 Iranian Revolutionary Guards Corps serve as good examples among ASAs, despite the obvious differences. Among ANSAs, IS stands out. Hence, meticulously unpacking the combat-relevant capacities and capabilities of IS and other ANSAs engenders and will keep on engendering new insights, research questions and hypothesis for scholars, policy/military analysts and decision-makers. Overall, it is what the agents *qualitatively* do with whatever they have – not necessarily *the quantity and the quality* of whatever they have – that matters more in combat performance. Given that knowhow is transferrable, the developments and enhancements in ANSAs' combat performance are likely to continue, with or without state-sponsorship or favourable environments and conditions. Furthermore, with its tactical innovations and strategic shifts, IS has inadvertently authored an upgraded textbook for others. The beneficiaries may include weaker ANSAs combating either stronger ASAs, state-sponsored ANSAs, maximalist and less-maximalist ANSAs and perhaps some of the ASAs of non-democracies. While IS failed in the 'art of the conduct of war', what it has exhibited in the 'art of fighting' can and probably will transform our understandings of insurgent warfare for many years to come.

Notes

1. 'President Trump on Syria: I'm not siding with anybody', *C-Span*, 7/10/2019, Accessed on 22/01/2020, at: shorturl.at/mDFNX
2. 'Iraq is in "Daily Battles" against the State Organisation in Kirkuk', [in Arabic] *al-Arab*, 19/04/2020, Accessed on 19/04/2020, at: shorturl.at/rwAF7
3. '*Hasad al-Ajnad* [Soldiers' Harvest]', *al-Naba'*, issue no. 224, Hijri-dated 10 Rajab 1441 [5 March 2020], p. 12.
4. Mu'ayad Bajis, 'For the First time . . . the State Organisation Reveals its Structure in Details', [in Arabic] *Arabi 21*, 06/07/2016, Accessed on 07/07/2017, at: shorturl.at/bkJS5
5. 'More than 1,000 Operations in 4 Months Kills 135 Terrorists', [in Arabic] *Iraqi24*, 23/04/2020, Accessed on 23/04/2020, at: shorturl.at/kltE3
6. These ten military sectors are Anbar, Salahuddin, Badiya, Jazeera, Nineveh, Tigris, Diyala, Kirkuk, Baghdad/North-Baghdad and the South Sector.
7. In Arabic, 'the comprehensive operation' is *al-'Amaliyya al-Shamilah* and 'the failed operation' is *al-'Amaliyya al-Fashilah*. SP makes use of this type of mockery-messaging for propaganda purposes.

8. Hugo Kaaman, 'Factories of Destruction', *Hugo Kaaman Open Source on SVBIEDs*, 31/03/2020, Accessed on 10/04/2020, at: shorturl.at/auzHT

9. See the section entitled 'Definitions and Terminology' in Introduction.

10. One example is that of Saddam al-Jamal, the ideologically uncommitted, barely literate smuggler who became the notorious *emir* of al-Bukamal in Deir Ezzor under IS. See: Hamzeh al-Mustafa, 'The Islamic State in Syria: Origins and the Environment', [in Arabic] in: Azmi Bishara (editor), *The State Organisation (Acronym ISIS): Foundation, Discourse and Practice*, [in Arabic] vol. 2, Doha: the Arab Centre for Research and Policy Studies, pp. 324–33.

11. As explained earlier – depending on the sub-state administrative delineations – 'foreign' and 'local' have different meanings. In al-Bukamal of Syria (Deir Ezzor governorate), a fighter from al-Qa'im of Iraq (Anbar governorate) would be considered less 'foreign' and more 'local' than a fighter from Damascus. In Sheikh Zuweid of Egypt, a fighter from Palestinian Rafah would be considered more of a 'local' than any fighter from the Nile Valley's cities and towns.

12. State-institutionalised military training and education.

13. Non-state ad-hoc or institutionalised military training, education and experiences.

14. Notable exceptions in terms of combat performance include ANSAs such as Hizbullah, Ansarullah (Houthis), Hamas and others. But as explained in the first chapter, their combat performance and overall effectiveness can be explained by state-sponsorship, local and regional support, and other variables reviewed in that chapter. *Hay'at Tahrir al-Sham* (Levant Liberation Organisation or HTS) is an example that supports the general argument. It uses similar tactics to that of IS (albeit with a different overall strategy) and some of its members, including its leader, came out of ISI and ISIS.

15. See for example the section entitled 'Striking from the Ashes: An Overview of IS Military Build-Up in Iraq' in Chapter 1.

16. In March 2008, al-Muhajir dissolved all non-IED tactical units, including sniping units and guerrilla formations. He focused all ISI's resources on building IED-tactical units. See the section entitled 'Striking from the Ashes: An Overview of IS Military Build-Up in Iraq' in Chapter 2.

17. See Chapter 2, footnote 76.

18. Al-Tunsi disappeared during the battle of al-Jabal in Deir Ezzor in December 2014. In 2015, allegations surfaced that he was killed due to an 'order of battle' dispute between him (as the *wali* of Deir Ezzor) and the military *emir* of Deir Ezzor. There were also ideological, tribal and ethnic elements causing some of these fractures. On that see for example: Aymenn al-Tamimi, 'Dissent in the Islamic State: Abu al-Faruq al-Masri's Message on the Menhaj', *Combating Terrorism Center*, 31/10/2016, Accessed on 11/11/2016, at: shorturl.at/duX34.

19. 'Interview: It will be a Fire that Burns the Cross and its People in Raqqah', *Rumiyah*, issue 12, 6 August 2017, p. 33.

20. See the section entitled 'Striking from the Ashes: An Overview of IS Military Build-Up in Iraq' in Chapter 2.

21. Some of these ANSAs formed alliances to diversify their combat experiences and boost their manpower, such as the cases of the DMSC with the Egyptian *al-Murbitun* (Border Fighters) organisation and of Syrian rebel organisations with the brigade-sized, about 4,000-strong Turkistan Islamic Party in Syria (ethnic Uyghur fighters).

22. As a reminder, the first hypothesis is 'IS tactically and operationally endures or expands due to successful shifts between conventional, guerrilla and terrorism strategies/ways of warfare. The second hypothesis is 'IS orchestrates tactical and operational military upsets against stronger enemies due to innovative combinations of tactics associated with conventional, guerrilla and terrorism ways of warfare'.

23. For samples of this behaviour see the section entitled 'Information as Power' in Norvell B. De Atkine, 'Why Arabs Lose Wars?', *Middle East Quarterly* (December 1999), Accessed on 19/02/2017, at: shorturl.at/dyJPV; also see Kenneth Pollack, *Arabs at War* (Nebraska: Nebraska University Press, 2002), p. 307; p. 717.

24. It is a 'cult' or a 'dogma' because it is closer to a belief-system in counterattacking as a 'sacred duty' than any rational military planning or costs-benefits calculations.

25. 'Interview: It will be a Fire that Burns the Cross and its People in Raqqah', *Rumiyah*, issue 12, 6 August 2017, pp. 32–5; Muhammad Al-Husan, Commander in the 166th Battalion, Interview in: '*Al-Rimal al-Mutaharika*: Sirte', [In Arabic], Documentary, 25/2/2018, Accessed on 11/12/2019, at: shorturl. at/zCDOW. For the details, see Chapter 2 (section entitled 'How ISIS/IS Fights in Syria: Battlefronts Analysis') and Chapter 3 (section entitled 'How IS Fights in Libya').

26. Even in terms of quality, weapons of mass destruction remain very far from the organisation.

27. Bernard Montgomery, *A History of Warfare* (London: Collins, 1968), p. 103.

28. As shown in Chapter 5, basic infantry and outdated air-defence tactics were used in the battle of Sheikh Zuweid in 2015. SP, however, does not rely on conventional warfare, given its very limited resources.

29. 'Interview with Abu Mus'ab al-Zarqawi', [Arabic] *Minbar al-Tawhid wa al-Jihad* (Hijri-year 1427 or 2006), Accessed on 15/4/2019, at: https://bit. ly/2xzgJFh

30. Former Commander in *al-Bunyan al-Marsus* Operation, Interview conducted by author, Istanbul, 12/11/2018.

31. Werner Wider, '*Auftragstaktik* and *Innere Fuhrung*: Trademarks of German Leadership', *Military Review* (September–October 2002), pp. 3–4.

32. Jack Welsh (1981) quoted in Dominik Thoma, *Moltke Meets Confucius: The Possibility of Mission Command in China* (Marburg: Tectum Verlag, 2016), p. 1.

33. For an in-depth comparative analysis see Eitan Shamir, *Transforming Command: The Pursuit of Mission Command in the U.S., British, and Israeli Armies* (Stanford: Standard University Press, 2011).

34. Norvell B. De Atkine, 'Why Arabs Lose Wars?'; Omar Ashour, 'The Enigma of Endurance and Expansion', [in Arabic] *Policy Studies* (Doha: al-Jazeera Centre for Studies, March 2016), p. 8, accessed on 21 January 2020, at: shorturl.at/vBY25

Bibliography

Chapter 1

A. W. Former FSA Commander in Raqqa. Interview by author. Istanbul, 9 October 2016.

Abduljabbar, Falih, *The Caliphate State: Advancing towards the Past*, [in Arabic] Doha: Arab Center for Research and Policy Studies, 2017.

Abu Rumman, Mohamed, and Hasan Abu Haniya, *The Islamic State: The Sunni Crisis and the Struggle over Global Jihadism*, [in Arabic] Amman: Friedrich-Ebert, 2015.

Arreguín-Toft, Ivan, 'How the Weak Win Wars: A Theory of Asymmetric Conflict', *International Security*, vol. 26, no. 1 (2001), pp. 93–128.

Ashour, Omar (ed.), *Punching Above Weights: Combat Capacities of Armed Non-State Actors*, Doha: Arab Centre for Research and Policy Studies, forthcoming in 2021.

Ashour, Omar (ed.), *Bullets to Ballots: Collective De-Radicalisation of Armed Movements*, Edinburgh: Edinburgh University Press, forthcoming in 2021.

Ashour, Omar, 'Ballots to Bullets: Transformations from Armed to Unarmed Activism', *Strategic Papers*, Doha: Arab Centre for Research and Policy Studies, 2018.

Ashour, Omar, 'Post-Jihadism and Ideological De-Radicalization', in: Zaheer Kazimi and Jeevan Doel (eds), *Contextualizing Jihadi Ideologies*, New York: Columbia University Press, 2011.

Ashour, Omar, *The De-Radicalization of Jihadists: Transforming Armed Islamists Movements*, London: Routledge, 2009.

Arab Opinion Index, 'The 2017–2018 Arab Opinion Index: Main Results in Brief', *Arab Centre for Research and Policy Studies*, at: https://bit.ly/38xi9OF, pp. 34–6.

Bajis, Mu'ayad, 'IS Executes *Sharia* Judge for Extreme Takfirism', [in Arabic] *Arabi21*, 8/8/2015.

Barrett, Richard, 'Beyond the Caliphate: Foreign Fighters and Threat of Returnees', *Soufan Center* (October 2007), pp. 1–40.

Bayo, Alberto, *150 Questions for a Guerrilla*, Denver: Cypress, 1963.

Berger, J. M., and Jessica Stern, *ISIS: The State of Terror*, New York: Ecco Press, 2015.

Biddle, Stephen, *Military Power: Explaining Victory and Defeat in Modern Battle*, Princeton: Princeton University Press, 2006.

Bishara, Azmi, *The State Organisation (Acronym ISIS): A General Framework and a Contribution to Help Understand the Phenomenon* – Vol. 1. [in Arabic], Doha: the Arab Centre for Research and Policy Studies, 2018.

Bishara, Azmi (ed.), *The State Organisation (Acronym ISIS): Foundation, Discourse and Practice* – Vol. 2. [in Arabic], Doha: Arab Centre for Research and Policy Studies, 2018.

Boulding, Kenneth, *Conflict and Defense: A General Theory*, New York: Harper, 1962.

Brooks, Risa, 'Making Military Might: Why Do States Fail and Succeed?' *International Security*, vol. 28, no. 2 (2003), pp. 149–91.

Byman, Daniel. et al, *Trends in Outside Support for Insurgent Movements*, Santa Monica: RAND Corporation, 2001.

Cochran, Kathryn M. and Stephen B. Long, 'Measuring Military Effectiveness: Calculating Casualty Loss-Exchange Ratios for Multilateral Wars, 1816–1990', *International Interactions*, vol. 43, no. 6 (2017), pp. 1019–40.

Condra, Luke and Jacob Shapiro. "Who Takes the Blame: The Strategic Effects of Collateral Damage." *American Journal of Political Science*, vol. 56, no. 1, pp. 167–89.

Connable, Ben, and Martin C. Libicki, *How Insurgencies End*, Arlington: Rand Publications, 2010.

Cordesman, Anthony, *Arab-Israeli Military Forces in an Era of Asymmetric Warfare*, London: Praeger, 2006.

David, Steven R, 'Why the Third World Matters', *International Security*, vol. 14, no. 1 (Summer 1989), pp. 50–85.

Downes, Alexander, *Targeting Civilians in War*, Ithaca, NY: Cornell University Press, 2008.

Dreyfuss, Bob, 'How the US War in Afghanistan Fuelled the Taliban Insurgency', *The Nation*, 18/9/2013.

Engelhardt, Michael, 'Democracies, Dictatorships, and Counterinsurgency: Does Regime Type Really Matter?" *Conflict Quarterly*, no. 12 (1992), pp. 52–63.

Fearon, James, and David D. Laitin, 'Ethnicity, Insurgency, and Civil War', *American Political Science Review*, vol. 97, no. 1 (2003), pp. 75–90.

Filiu, Jean-Pierre, *From Deep State to Islamic State: The Arab Counter-Revolution and Its Jihadist Legacy*, New York: Oxford University Press, 2005.

Former Iraqi Military Intelligence General. Interview by author, London, August 2017.

Freedman, Lawrence, *Strategy a History*, Oxford: Oxford University Press, 2013.

Galula, David, *Counterinsurgency Warfare: Theory and Practice*, Westport, CT: Praeger, 1964.

Gemtansky, Anna, 'You Can't Win If You Don't Fight', *Journal of Conflict Resolution*, vol. 57, no. 4 (2012), pp. 709–32.

Gerges, Fawaz, *ISIS: A History*, Princeton: Princeton University Press, 2016.

Guevara, Ernesto, *Guerrilla Warfare*, North Melbourne: Ocean Press, 1961.

Hafez, Mohammed, *Why Muslims Rebel? Repression and Resistance in the Islamic World*, London: Lynne Rienner, 2003.

Hassan, Hassan, 'The Sectarianism of the Islamic State', *Carnegie Papers*, June 2016.

Hegghammer, Thomas, and Petter Nesser, 'Assessing the Islamic State Commitment to Attacking the West', *Perspectives on Terrorism*, vol. 9, no. 4 (2015).

Hoffman, Bruce, *Inside Terrorism*, New York: Columbia University Press, 2006.

Hume, Tim, 'Battle for Mosul: How ISIS Is Fighting Back', *CNN*, 25/10/2016.

al-Ibrahim, Bader, 'ISIS, Wahhabism and Takfir', *Contemporary Arab Affairs*, vol. 8, no. 3 (2015), pp. 408–15.

Johnston, Patrick B, 'The Geography of Insurgent Organisation and its Consequences for Civil Wars: Evidence from Liberia and Sierra Leone', *Security Studies*, vol. 11 (2008), pp. 107–37.

Jones, Seth, *Waging Insurgent Warfare*, Oxford: Oxford University press, 2017.

Jones, Seth, and Patrick Johnston, 'The Future of Insurgency', *Studies in Conflict and Terrorism*, vol. 36, no. 1 (2013), pp. 1–25.

Kalyvas, Stathis N, and Laia Balcells, 'International System and Technologies of Rebellion: How the End of the Cold War Shaped Internal Conflict', *American Political Science Review*, vol. 104, no. 3 (August 2010), pp. 415–29.

Kalyvas, Stathis, *The Logic of Violence in Civil War*, New York: Cambridge University Press, 2006.

Keating, Tom, 'The Importance of Financing in Enabling and Sustaining the Conflict in Syria (and Beyond)' , *Perspectives on Terrorism*, vol. 8, no. 4 (2014).

Keefer, Philip, 'Insurgency and Credible Commitment in Autocracies and Democracies', *The World Bank Economic Review*, vol. 22, no. 1 (2008), pp. 33–61.

Kevlihan, Rob, Aid, *Insurgencies and Conflict Transformation: When Greed is Good*, London: Routledge, 2012.

Kilcullen, David, *Out of the Mountains: The Coming Age of the Urban Guerrilla*, New York: Oxford University Press, 2013.

Kilcullen, David, *The Accidental Guerrilla: Fighting Small Wars in the Midst of a Big One*, Oxford: Oxford University Press, 2009.

Lagon, Mark P, 'The International System and the Reagan Doctrine: Can Realism Explain Aid to 'Freedom Fighters'?', *British Journal of Political Science*, vol. 22, no. 1 (January 1992), pp. 39–70.

Lansdale, Edward G, 'Vietnam: Do We Understand Revolution?', *Foreign Affairs* vol. 43, no. 1 (October 1964), pp. 75–86.

Lansdale, Edward G., *In the Midst of Wars: An American's Mission to Southeast Asia*, New York: Harper and Row, 1972.

Laqueur, Walter, *Guerrilla Warfare: A Historical and Critical Study*, New Brunswick, NJ: Transaction, 1998.

Lawrence, T.E, 'Science of Guerrilla Warfare', in: Malcolm Brown and T. E. Lawrence (ed.), *War and Peace: An Anthology of the Military Writings of Lawrence of Arabia*, London: Greenhill Books, 2005.

Layada, Abdul Haqq, 'Layada to *al-Hayat*: I Founded the Armed Group to Defend the Algerian People and Not to Kill them', [in Arabic] Interview by Camille, al-Tawil, *al-Hayat*, 6/6/2007, p. 6.

Luttwak, Edward, 'The Operational-Level of War', *International Security*, vol. 5, no. 3 (Winter 1980–1), pp. 61–79.

Lyall, Jason, and Isaiah Wilson, 'Rage against the Machines: Explaining Outcomes in Counterinsurgency Wars', *International Organisation*, vol. 63, no. 1 (2009), pp. 67–106.

Lyall, Jason, 'Do Democracies Make Inferior Counterinsurgents? Reassessing Democracy's Impact on War Duration and Outcomes', *International Organization*, vol. 64, no. 1 (2010), pp. 167–92.

Macaulay, Neil, 'The Cuban Rebel Army: A Numerical Survey', *The Hispanic American Historical Review*, vol. 58, no. 2 (1978), pp. 284–95.

Mack, Andrew, 'Why Big Nations Lose Small Wars: The Politics of Asymmetric Conflict', *World Politics*, vol. 27, no. 2 (1975), pp. 175–200.

McCants, William, *ISIS Apocalypse*, London: Picador, 2016.

McColl, Robert, 'The Insurgent State: Territorial Bases of Revolution', *Annals of the Association of American Geographers*, vol. 59, no. 4 (1969), pp. 613–31.

Michael, Doyle W., and Nicholas Sambanis, *Making War and Building Peace*, Princeton, NJ: Princeton University Press, 2006.

Millett, Allan. et al, 'The Effectiveness of Military Organizations', *International Security*, vol. 11, no. 1, 1986, pp. 37–71.

Milton, Daniel. Bryan Price and Muhammad al-'Ubaydi, 'The Islamic State in Iraq and the Levant: More than Just a June Surprise', *CTC sentinel*, vol. 7, no. 6 (June 2014).

Nagl, John, *Kinfe Fights: A Memoir of Modern War in Theory and Practice*, London: Penguin, 2014.

Nesser, Petter, *Islamist Terrorism in Europe: A History*, Oxford: Oxford University Press, 2018.

Pape, Robert A., and James K. Feldman, *Cutting the Fuse: The Explosion of Global Suicide Terrorism and How to Stop It*, Chicago: University of Chicago Press, 2010.

Pape, Robert A., *Dying to Win: The Strategic Logic of Suicide Terrorism*, New York: Random House, 2005.

Perry, Mark, 'How Iraq's Army Could Beat ISIS in Mosul, But Lose the Country', *Politico*, 15/12/2016.

Petraeus, David, James F. Amos and John A. Nagl, *The U.S. Army/Marine Corps Counterinsurgency Field Manual*, Chicago: University of Chicago Press, 2007.

Pike, Douglas, *The Organization and Techniques of the National Liberation Front of South Vietnam*, Boston: MIT Press, 1966.

Pollack, Kenneth, *Arabs at War: Military Effectiveness, 1948–1991*, Lincoln: University of Nebraska Press, 2004.

Pollack, Kenneth, *Armies of Sands*, Oxford: Oxford University Press, 2019.

Powell, Jonathan, *Talking to Terrorists*, London: Vintage, 2015.

Al-Rasheed, Madawi, *A History of Saudi Arabia*, Cambridge: Cambridge University Press, 2010.

Remnick, David, 'Going the Distance: On and Off the Road with Barack Obama', *The New Yorker*, 20/1/2014.

Rosenau, William, 'The Kennedy Administration, US Foreign Internal Security Assistance and the Challenge of "Subterranean War," 1961–63', *Small Wars and Insurgencies*, vol. 14, no. 3 (Autumn 2003), pp. 65–99.

Salehyan, Idean, Kristian Skrede Gleditsch and David E. Cunningham, 'Explaining External Support for Insurgent Groups', *International Organization*, no. 65 (Fall 2011), pp. 709–44.

Sanín, Francisco Gutiérrez, and Elisabeth Jean Wood, 'Ideology in Civil Wars', *Journal of Peace Research*. Vol. 51, no. 2 (2014), pp. 213–26.

Schutte, Sebastian, 'Geography, Outcome, and Casualties: A Unified Model of Insurgency', *Journal of Conflict Resolution* (March 2014), pp. 1–28.

Scuitto, Jim. Et al, 'ISIS can "muster" between 20,000 and 31,500 fighters, CIA says', *CNN*, 12/9/2014.

Seig, Hans, 'How the Transformation of Military Power Leads to Increasing Asymmetries in Warfare?', *Armed Forces and Society*, vol. 40, no. 2 (2014), pp. 332–56.

Shaikhy, Umar, 'The Emir of al-Akhdariya Uncovers the Secrets of the Mountain', Interview by Camille al-Tawil [in Arabic]. *Al-Hayat*, 7/6/2007, p. 10.

Staniland, Paul, 'Organizing Insurgency: Networks, Resources, and Rebellion in South Asia', *International Security*, vol. 37, no. 1 (Summer 2012), pp. 142–77

Souaïdia, Habib, *La Sale Guerre*, Paris: La Découverte, 2001.

Al-Suri, Abu Mus'ab, *The Summary of My Testimony on the Jihad in Algeria*, [in Arabic] Afghanistan: No Publisher, No Date.

El-Tahawy, Mona, 'Lives Torn Apart', *The Guardian*, 20/10/1999.

Al-Tawil, Camille, *al-Qa'ida and Its Sisters*, [in Arabic] Beirut: Dar al-Saqi, 2007.

Thompson, R., *Defeating Communist Insurgency: The Lessons of Malaya and Vietnam*, Westport: Praeger, 1966.

Tse-tung, Mao, *On Guerrilla Warfare*, Champaign: University of Illinois, 1937.

Walter, Barbara, 'Why Bad Governance Lead to Repeat Civil Wars', *Journal of Conflict Resolution*, vol. 59, no. 7 (2014), pp. 1242–72.

Waltz, Kenneth N, *Theory of International Politics*, New York: McGraw-Hill, 1979.

Warrick, Joby, *Black Flags: The Rise of ISIS*, London: Corgi, 2016.

Weiss, Michael, and Hassan Hassan, *Inside Army of Terror*, New York: Regan Arts, 2015.

Wells, Matthew, 'Casualties, Regime Type and the Outcomes of Wars of Occupation', *Conflict Management and Peace Sciences*, vol. 33, no. 5 (2016), pp. 469–90.

Wolff, Terry, Conversation with author, Prague, 10 February 2017.

'Baghdad Falls to US Forces', *BBC News*, 9/4/2003.

'The Taliban Are Forced Out of Afghanistan', *BBC History*, 18/7/2018.

Chapter 2

al-Abidi, Manaf, 'The Story of the Fall of Ramadi', [in Arabic] *al-Sharq al-Awsat*, 26/6/2015.

Abu Rajab, Former FSA Fighter who fought ISIS/IS in Idlib, Aleppo and Raqqa. Interview by author, Istanbul, 9 October 2016.

Abdul-Zahra, Qassim, and Lara Jakes, 'Al Qaeda Iraq Strength Musters', *Associated Press*, 10/10/2012.

al-Ansary, Khalid, and Ali Adeeb, 'Most Tribes in Anbar Agree to Unite Against Insurgents', *The New York Times*, 19/09/2018.

Arango, Tim, 'Dozens Killed Across Iraq as Sunnis Escalate Protests Against Government', *The New York Times*, 24/4/2013.

Ashour, Omar, 'Viewpoint: How Islamic State is Managing to Survive', *BBC News*, 14/12/2015.

Barfi, Barack, 'The Military Doctrine of the Islamic State and the Limits of Ba'athist Influence', *CTC Sentinel*, vol. 9, no. 2 (February 2016).

Biddle, Stephen. Jeffrey A. Friedman and Jacob N. Shapiro, 'Testing the Surge: Why Did Violence Decline in Iraq in 2007?', *International Security*. vol. 37, no. 1 (Summer 2012), pp.7–40.

Blake, Paul, 'Ramadi Assault: How a Small Change in Tactics helped Iraqi forces', *BBC*, 22/12/2015.

Bokel, John, 'IEDs in Asymmetric Warfare', *Military Technology*, vol. 31, no. 10 (October 2007).

Bulos, Nabih, 'Islamic State Has Been Cranking Out Car Bombs on an Industrial Scale for the Battle of Mosul', *New York Times*, 25/2/2017.

Coles, Isabel, John Walcott and Maher Chmaytelli, 'Islamic State Leader Baghdadi Abandons Mosul Fight to Field Commanders, U.S. and Iraqi Sources Say', *Reuters*, 8/3/2017.

al-Falih, Rashid, Commander of the Popular Mobilization Forces in al-Anbar. Interview by Qusai Shafiq, *Al-Ahd Channel*, Iraq, 27/8/2017.

Firik, Kemal, 'ISIS's Weapon Inventory Grows', *Daily Sabah Mideast*, 2/7/2014.

Flood, Derek Henry, 'From Caliphate to Caves: The Islamic State's Asymmetric War in Northern Iraq', *CTC Sentinel*, vol. 11, no. 8 (September 2018).

Former Lieutenant-General in the Iraqi Army. Interview by the author. Doha, March 2019.

Former Major-General in the Iraqi Army. Interview by the author. Istanbul, 7 February 2019.

Former Peshmerga Commander. Interview by the author. Prague, February 2017.

Grier, Peter, 'April 15, 1953', *Airforce Magazine* (June 2011), p. 54.

al-Hamed, Raed, 'The Battle of Ramadi: The Coalition's Gains and ISIS' Tactics', [in Arabic] *Aljazeera Centre for Studies*, 13/2/2016.

al-Hashimi, Hesham, Former Iraqi government security advisor. Interview by Qusai Shafiq, *Al-Ahd Channel*, Iraq, 27/8/2017.

Hassan, Falih et al, 'Celebrating Victory Over ISIS, Iraqi Leader Looks to Next Battles', *The New York Times*, 30/12/2015.

Hume, Tim, 'Battle for Mosul: How ISIS Is Fighting to Keep its Iraqi Stronghold', *CNN*, 25/10/2016.

International Crisis Group Iraq, 'Falluja's Faustian Bargain', *Middle East Report*, no. 150, 28/4/2014.

Ismay, John, Thomas Gibbons-Neff and C. J. Chivers, 'How ISIS Produced Its Cruel Arsenal on an Industrial Scale', *New York Times*, 10/12/2017.

Issam, Wael, 'ISIS Gains Control of Fallujah after Capturing Military Council Officers', [in Arabic] *Al-Quds Al-Arabi*, 28/6/2014.

al-Issawi, Ghassan, Spokesperson for the Anti-IS Tribal Coalition in Ramadi. Interview by Qusai Shafiq. *Al-Ahd Channel*, Iraq, 27/8/2017.

Janabi, Omar, 'Tunnels of Death . . . The New Weapon of ISIS', [Arabic] *al-Khaleej Online*, 16/03/2015.

Jones, Susan, 'Ramadi Liberated, But Booby Trapped', *CNS News*, 11/2/2016.

Joscelyn, Thomas, 'Analysis: Islamic State Claims Historically High Number of Suicide Attacks in 2016', *Long War Journal*, 3/1/2017.

Kaaman, Hugo, 'Islamic State Statistics on its SVBIED Use from Late 2015 through 2017, Including the Battle of Mosul', *Hugo Kaaman Open Source on SVBIEDs*, 18/8/2018.

Kamaan, Hugo, 'Islamic State SVBIED Development since 2014', Paper presented at the Annual Conference of the Strategic Studies Unit entitled 'Militias and Armies: Developments of Combat Capacities of Armed Non-State and State Actors', *Arab Centre for Research and Policy Studies* (ACRPS), Doha, 24 February 2020.

Kilcullen, David, *Blood Year*, Oxford: Oxford University Press, 2016.

Knickerbocker, Brad, 'Relentless Toll to US Troops of Roadside Bombs: The IED Has Caused over a Third of the 3,000 American GI Deaths in Iraq', *Christian Science Monitor*, 2/1/2007.

Lake, Eli, Jamie Dettmer and Nadette De Visser, 'Iraq's Terrorists are Becoming a Full-Blown Army', *Daily Beast*, 6/11/2014.

Larter, David B, 'SOCOM Commander: Armed ISIS Drones Were 2016's "Most Daunting Problem"', *Defense News*, 16/5/2017.

Lewis, Jessica, 'Al Qaeda in Iraq is Resurgent', *Middle East Security Report*, no. 14 (September 2013), p. 8.

Losey, Stephen, 'Airstrikes Against ISIS Hit all-time High', *Airforce Times*, 13/9/20017.

Losey, Stephen, 'With 500 Bombs a Week, Mosul Airstrikes Mark "the Most Kinetic" Phase of ISIS Air War So Far', *Airforce Times*, 28/3/2017.

Malkasian, Carter, 'Anbar's Illusions: The Failure of Iraq's Success Story', *Foreign Affairs*, 24/6/2017.

Marcus, Jonathan, 'Mosul: Have Combat Changes Increased Civilian Casualties?' *BBC*, 29/3/2017.

Martin, Patrick, Genevieve Casagrande and Jessica Lewis McFate, 'ISIS Captures Ramadi', *Institute for The Study of War*, 18/5/2015.

McCants, William, *The ISIS Apocalypse: The History, Strategy and Doomsday Vision of the Islamic State*, New York: St Martin's Press, 2015.

Namaa, Kamal, 'Iraqi Militants Kill at Least 18 Soldiers, Including Commander', *Reuters*, 21/12/2013.

Office of the Inspector General, 'Operation Inherent Resolve and Other Overseas Contingency Operations: October 1, 2018 – December 31, 2018', *US Department of Defense*, 4/2/2019, p. 21.

Office of the Inspector General, 'Overseas Contingency Operations: Operation Inherent Resolve and Operation Pacific Eagle- Philippines Report to the United States Congress: 1 April 2018–30 June 2018', *US Department of Defense*, 6/8/2018.

Parker, Ned, Isabel Coles and Raheem Salman, 'Special Report: How Mosul Fell - An Iraqi General Disputes Baghdad's Story', *Reuters*, 14/10/2014.

Pawlyk, Oriana, 'Diverting to Fallujah from Syrian Town was Right Call to Target ISIS, General Says', *Airforce Times*, 15/7/2016.

Perry, Mark, 'How Iraq's Army Could Beat ISIS in Mosul But Lose the Country', *Politico*, 15/12/2016.

Rassler, Don, 'The Islamic State and Drones: Supply, Scale and Future Threats', *Combatting Terrorism Center at West Point* (July 2018).

Rassler, Don, Muhammad Al-`Ubaydi and Vera Mironova, 'The Islamic State's Drone Documents: Management, Acquisitions, and DIY Tradecraft', *Combatting Terrorism Centre at West Point*, 31/1/2017.

Roggio, Bill and Caleb Weiss, 'Islamic State Photos Highlight Group's Grip on Ramadi', *Long War Journal*, 16/10/ 2014.

Roggio, Bill, 'ISIS Parades on Outskirts of Baghdad', *The Long War Journal*, 1/4/2014.

Said, Haider, 'The Way to the Fall of Mosul', [in Arabic] *Siyasat Arabiya*, no. 10 (September 2014).

Schmidt, Michael, 'Suicide Bombs in Iraq Have Killed 12,000 Civilians, Study Says', *New York Times*, 2/11/2011.

Schmidt, Michael and Eric Schmitt, 'Pentagon Confronts a New Threat from ISIS: Exploding Drones', *The New York Times*, 12/10/2016.

Sfanson, Berguen, 'Interesting Facts about the Fall of Mosul to ISIS a Year Ago', [in Arabic] Trans. Rim Nejmi, *Deutsche Welle*, 1/6/2015.

Shanker, Thom, 'Al-Qaeda Leaders in Iraq Neutralised, U.S. Says', *The New York Times*, 4/6/2010.

Smisim, Issa, '*al-Inghimassiyun*: The Striking Force of Jihadist Organizations', [in Arabic] *al-Arabi al-Jadid*, 22/12/2014.

Sullivan, Ben, 'The Islamic State Conducted Hundreds of Drone Strikes in Less Than a Month', *Vice*, 21/2/2017.

Triebert, Christiaan, 'An Open Source Analysis of the Fallujah 'Convoy Massacre'(s)', *Bellingcat*, 6/7/2016.

Warren, Steve, 'Department of Defense Press Briefing by Col. Warren via Teleconference from Baghdad, Iraq', *US Department of Defence*, 29/12/2015.

Waters, Nick, 'Types of Islamic State Drone Bombs and Where to Find Them', *Bellingcat*, 24/5/2017.

Whiteside, Craig, and Vera Mironova, 'Adaptation and Innovation with an Urban Twist Changes to Suicide Tactics in the Battle for Mosul', *Military Review* (November–December 2017), pp. 78–85.

Wilayat al-Fallujah, '14 Martyrdom Operations', [in Arabic] *Al-Naba'*, no. 36, 24/5/2016, p. 6.

Witness from al-Rabee neighbourhood in Western Mosul. Interview by the author, 8 January 2019.

Witty, David M, 'Iraq's Post-2014 Counter-Terrorism Service', *Washington Institute for Near East Policy*, October 2018.

Youssef, Nancy A, and Shane Harris, 'How ISIS Actually Lost Ramadi', *Daily Beast*, 30/12/2015.

'824 Prisoners Freed by Force since Al-Baghdadi Announced 'Breaking the Walls' Campaign', [in Arabic] *Al-Hayat*, 25/7/2013.

'al-Aan News Reports on ISIS's Use of Tunnels to Infiltrate the City of Ramadi in Anbar', *al-Aan News*, al-Aan TV, 26/2/2015.

'Anbar Governorate: 85 Bridges were Destroyed by ISIS and We Need 160 Million Dinars to Rebuild them', *Al Mada Press*, 25/10/2015.

'Announcement of ISIS in Mosul', [in Arabic] *Al Jazeera*, 12/6/2015.

'Department of Defense Briefing by General Townsend via Telephone from Baghdad, Iraq', *US Department of Defense*, 28/3/2017.

'Documents of the Second Man Answers: How ISIS Controlled al-Mosul', [in Arabic] *Qasioun*, 5/6/2018.

'Explosives Found in Mutaibijah Enough to Booby-Trap 50 Cars', [in Arabic] *Al-Sumaria TV*, 18/12/2017.

'Foreign Fighters Continue to Join ISIS in Syria, US Joint Chiefs Chair Says', *Defense Post*, 16/10/ 2018.

'al-Furqan Media Presents a News Bulletin from Islamic State of Iraq and al-Sham: Harvest of Operations for the Year 1433 H in Iraq', *Jihadology*, 14/8/2013.

'al-Furqan Media Presents a New Video Message from the Islamic State: And They Gave Zakah', *Jihadology*, 17/6/2015.

'How Did Extremists Take Over One of Iraq's Biggest Cities in Just Five Days?' *Niqash*, 10/6/2014.

'al-*Inghimassiyun* ISIS's Most Lethal Weapon', [in Arabic] *al-Khalij Online*, 17/8/2014.

'Interview with Abu Mus'ab al-Zarqawi', [in Arabic] *Minbar al-Tawhid wa al-Jihad*, 2006.

'*Irak: L'opération pour Reprendre Mossoul des Mains de l'EI est Lancée*', Le Monde, 17/10/2016.

'Iraq: British Contractor Killed Clearing Mines in Ramadi', *Sky News*, 22/8/2016.

'Less Than 1,000 IS Fighters Remain in Iraq and Syria, Coalition Says', *Reuters*, 27/12/2017.

'New Magazine from the Islamic State of Iraq and al-Sham: al-Bina' Magazine', *Jihadology*, 31/3/2014.

'New Video Message from the Islamic State: "Then They Will Be Overcome – Wilayat al-Iraq, Al-Fallujah"', *Jihadology*, 2/6/2019.

'Press Conference by Special Presidential Envoy McGurk in Erbil, Iraq', *US Department of State*, 4/9/ 2017.

'Temporarily Conquering Cities: *Modus Operandi* for Holy Fighters', [in Arabic] *Al-Naba'*, no. 180, 3/5/2019, p. 9.

'The Investigation Regarding the Fall of Mosul Reveals the Responsibility of Maliki and Other High-Ranking Officials', [in Arabic] *Al-Watan News*, 19/8/2015.

'This is How ISIS Controlled al-Mosul', [in Arabic] *Akhbar al-Yawm*, 12/6/2015.

'Translation of Old Al-Zarqawi Statement Says God's Law Must Rule Entire World', *Open Source Center*-GMP20061211281001, 6/12/2006, in: Ahmed Hashim, 'From al-Qaida Affiliate to The Rise of The Islamic Caliphate: The Evolution of the Islamic State of Iraq and Syria (ISIS)', *Policy Report, RSIS*, 12/2014.

'Weapons of the Islamic State: A Three-Year Investigation in Iraq and Syria', *Conflict Armament Research*, 12/2017, pp. 183–4.

Chapter 3

A. W. Former FSA Commander in Raqqa. Interview by author, Istanbul, 9 October 2016.

Aba Zeid, Ahmad, 'Abu Khaled Al-Suri: The First Generation in the Face of the Last Deviation', [in Arabic] *Zaman al-Wasl*, 26/2/2015.

Abouzeid, Rania, 'The Jihad Next Door: The Syrian Roots of Iraq's Newest Civil War', *Politico*, 23/6/2014.

Abu Al-Harith, former Army of Islam – North Commander. Interview by author, Gaziantep, 11 November 2018.

Abu Mansur, former FSA Fighter (Revolutionaries of Raqqa Brigade). Interview by author, Istanbul, 8 November 2018.

Abu Qutayba, former FSA Commander. Interview by author, Gaziantep, 12 November 2018.

Abu Rajab, former FSA Fighter who fought ISIS/IS in Idlib, Aleppo and Raqqa. Interview by author, Istanbul, 9 October 2016.

Abu Rumman, Mohamed, and Hasan Abu Haniya, *The Islamic State: The Sunni Crisis and the Struggle over Global Jihadism*, [in Arabic] Amman: Friedrich-Ebert, 2015.

Al-Arabi, Ahmed, 'ISIS Operates Air-Sorties in Raqqa', [in Arabic] *Aljazeera*, 28/10/2014.

Ali, Abdullah Suleiman, 'Showdown Begins between Syrian Army, Islamic State', *Al-Monitor*, 25/7/2014.

Amer, Obeida, 'Al Julani's Journey . . . from the Heart of "ISIS" to Cloning "Hizbullah"', [in Arabic] *Midan Aljazeera*, 3/1/2019.

Arquilla, John and David Ronfeldt, *Swarming and the Future of Conflict*, Santa Monica: RAND Corporation, 2000.

Ashour, Omar, 'Post-Jihadism: Libya and the Global Transformations of Armed Islamist Movements', *Terrorism and Political Violence*, vol. 13, no. 3 (June 2011), pp. 377–97.

Awad, Mayssa and James Andre, 'Exclusive: IS Group's Armoured Drones Attack from the Skies in Battle for Raqqa', *France 24*, 26/6/2017.

Bender, Jeremy, 'ISIS Just Looted Advanced Weaponry from a Crucial Assad Regime Airbase in Syria', *Business Insider*, 25/8/2014.

Bishara, Azmi (ed.), *The State Organisation (Acronym ISIS): Foundation, Discourse and Practice* – Vol. 2 [in Arabic], Doha: Arab Centre for Research and Policy Studies, 2018.

Darwich, Ibrahim, 'The Story of the Fall of Raqqa in the Hands of Jihadists . . . Betrayals, Executions and Cruel Practices', *Al-Quds Al-Arabi*, 16/6/2015.

Dillon, Ryan, and Eric J. Pahon, 'Department of Defense Press Briefing by Colonel Dillon via teleconference from Baghdad, Iraq', *US Department of Defense*, 3/8/2017.

Draper, Lucy, 'ISIS Controls over 50% of Syria After Taking Palmyra', *Newsweek*, 21/5/2015.

Eleftheriou-Smith, Loulla-Mae, 'Haji Bakr: Former Saddam Hussein Spy is Mastermind behind ISIS Takeover of Northern Syria and Push into Iraq, Report Claims Documents Uncovered by Der Spiegel Claim', *Independent*, 20/4/ 2015.

Evans, Dominic and Orhan Coskun, 'Defector Says Thousands of Islamic State Fighters Left Raqqa in Secret Deal', *Reuters*, 7/12/2017.

Former Army of Islam Commander. Interview by author, Istanbul, 7 November 2016.

Former Commander in the FSA and al-Jabha al-Shamiya. Levantine Front. Interview by author. Gaziantep, 13 November 2018.

Former Syrian Army Officer. Interview by author. Beirut, 6 December 2017

Francis, Ellen, and Issam Abdallah, 'Islamic State Deploys Car Bombs in Defense of Last Enclave', *Reuters*, 3/3/2019

Hamijou, Mohamed Manar, '"Al-Julani", "Al-Buwaydani", "Al-Shameer" and Dozens of Terrorists', [in Arabic] *Al Watan*, 11/12/2018.

Hassan, Hassan, 'Two Houses Divided: How Conflict in Syria Shaped the Future of Jihadism', *CTC Sentinel*, vol. 11, no. 9 (October 2018).

Hopkins, Alex, 'International Airstrikes and Civilian Casualty Claims in Iraq and Syria', *Airwars Reports*, 7/2017.

Horton, Alex, 'ISIS Fighters Booby-Trapped Corpses, Toys and a Teddy Bear in Besieged Raqqa', *The Washington Post*, 18/10/2017.

Ignatius, David, 'Al-Qaeda Affiliate Playing Larger Role in Syria Rebellion', *The Washington Post*, 30/11/2012.

Issam, Wael, 'Al-Quds Al-Arabi Narrates the Reasons for the Sudden Fall of Raqqa in the Hands of the Islamic State . . . the Relationship Between Jabhat al-Nusra and ISIS in the City . . . from Alliance to War', [in Arabic] *Al-Quds Al-Arabi*, 30/1/2014.

Issam, Wael, 'The Leader of the Al-Nusra Front for 'Al-Quds Al-Arabi': ISIS Fighters were Calling us on the Radio Saying "Ahrar Al-Sham Withdrew and Left You"', [in Arabic] *Al Quds Al Arabi*, 31/1/2014.

Issam, Wael, 'The Reasons for the Sudden Fall of Raqqa in the Hands of the Islamic State . . . the Relationship Between Jabhat al-Nusra and ISIS in the city . . . from Alliance to War', [in Arabic] *Safahat Souriyya*, 1/2/2014.

K., Saad, former SDF Fighter. Skype Interview by author, January 2019.

Kaaman, Hugo, 'From Hajin to Baghouz – Islamic State SVBIED Design and Use', *Hugo Kaaman Open Resource Research on SVBIEDs*, 3/8/2019.

Kamaan, Hugo, 'Islamic State SVBIED Development since 2014', Paper Given at the Annual Conference of the Strategic Studies Unit Entitled 'Militias and Armies: Developments of Combat Capacities of Armed Non-State and State Actors', *Arab Centre for Research and Policy Studies*, Doha, 24/2/2020.

Lister, Charles, *The Syrian Jihad*, London: Hurst, 2015.

Lister, Tim, 'Battle for Raqqa: Seven Things you Need to Know', *CNN*, 6/6/2017.

Al-Manara Al-Bayda Corporation for Media Production, 'Video: Announcing the Formation of the Support Front in the Levant to fight Bashar al-Assad', [in Arabic] *Dailymotion*, 12/2011.

Micallef, Joseph V, 'Sitrep Raqqa: The Geopolitics of Eastern Syria', *Military. Com*, 26/6/2017.

al-Mustapha, Hamzeh, 'Combat Performance of al-Nusra Front in Syrian Civil War', Paper presented at the Annual Conference of the Strategic Studies Unit entitled 'Militias and Armies: Developments of Combat Capacities of Armed Non-State and State Actors', *Arab Centre for Research and Policy Studies* (ACRPS), Doha, 23/2/2020.

Nasr, Wasim, 'Details of the Open Confrontation Between the 'Islamic State' and the Syrian Army', *France 24*, 28/1/2014.

Nasr, Wasim, 'The Repercussions of the fall of Tabqa Military Airport under ISIS', [in Arabic] *France 24*, 29/8/2014.

O'Connor, Tom, 'U.S. Made Secret Deal with Isis to Let Thousands of Fighters Flee Raqqa to Battle Assad in Syria, Former Ally Says', *Newsweek*, 12/8/17

Al Oolwani, Al Majeed, 'After Withdrawals and 'Betrayals'. . . Raqqa Rebels Regain 50% of the City', [in Arabic] *Orient Net*, 13/1/2014.

Reuter, Christoph, 'Secret Files Reveal the Structure of Islamic State', *Spiegel International*, 18/4/2015.

Roggio, Bill, and Caleb Weiss, 'More Jihadist Training Camps Identified in Iraq and Syria', *Long War Journal*, 23/11/2014.

Rota, Alessandro, 'From Teddy Bears to Bombs: the IEDs of Isis – in Pictures', *The Guardian*, 29/10/2016.

S., Jalal, former SDF Fighter. Skype Interview by author, December 2018.

S., Jawad, former Commander in *Liwa' al-Haqq* (Truth Brigade), Interview by author. Istanbul, 7 October 2016.

Scuitto, Jim. et al, 'ISIS Can "Muster" between 20,000 and 31,500 fighters, CIA says', *CNN*, 12/9/2014.

Seligman, Lara, 'In Overflowing Syrian Refugee Camps, Extremism Takes Root', *Foreign Policy*, 29/7/2019.

Shaheen, Kareem, 'ISIS "Controls 50% of Syria" after Seizing Historic City of Palmyra', *The Guardian*, 21/5/2015.

al-Shari', Saad. Interview by author. Gaziantep, 11 November 2018.

Smith, Martin, 'The Rise of ISIS', *Frontline*, 28/10/2014.

Snow, Shawn, 'These Marines in Syria Fired more Artillery than any Battalion
Since Vietnam', *Marine Corps Time*, 6/2/2018.

Sommerville, Quentin and Riam Dalati, 'Raqqa's Dirty Secret', *BBC News*,
13/11/2017.

South, Todd, '3rd Cavalry Regiment Soldiers are Firing Intense Artillery Missions
into Syria with Iraqi, French Allies', *Army Times*, 11/12/ 2018

al-Tamimi, Aymenn Jawad, 'The Islamic State's 'Revenge' Expedition for Abu
Bakr al-Baghdadi and Abu al-Hassan al-Muhajir: Data and Analysis', *aymen-
njawad.org*, 31/12/2019.

Townsend, Stephen J, 'Remarks by General Townsend in a Media Availability
in Baghdad, Iraq', *US Department of Defense*, 11/7/2017.

Van Wilgenburg, Wladimir, 'Raqqa in Ruins: Brutal Fight Against IS Leaves
City Destroyed', *Middle East Eye*, 28/9/2017

Westall, Sylvia, 'Hundreds Dead as Islamic State Seizes Syrian Air Base – Monitor',
Reuters, 24/8/2014

Whiteside, Craig, 'New Masters of Revolutionary Warfare: The Islamic State
Movement (2002–2016)', *Perspective on Terrorism*, vol 10, no. 4 (August
2016).

Wilayat al-Raqqa, 'Inghimassiyun and Martyrdom Operations', [in Arabic] *al-Naba'*,
no. 86, Hijri-dated 27 Ramadan 1438, p. 6.

Wilgenburg, Van Wilgenburg, 'Raqqa: IS Sows Chaos as Suicide and Tunnel
Attacks Blur Front Lines', *Middle East Eye*, 4/8/2017.

Zelin, Aaron Y, 'Wilayat al-Hawl: 'Remaining' and Incubating the Next Islamic
State Generation', *Policy notes*, *The Washington Institute for Near East Policy*
(October 2019).

'120 Fallen in a Wide-Scale Attack', [in Arabic] *al-Naba'*, no. 99, Hijri-dated
8 Muharram 1439, p. 14.

'After a Year and a Half Siege . . . the "Islamic State" Controls the 17th Division
of Raqqa', [in Arabic] *Alkhaleej Online*, 25/7/2014.

'After the Battle of Greater al-Raqqa Slowed, the Islamic State Uses Tunnels
Tactics and Besieges a Group of Fighters from "Operation Fury of the
Euphrates"', [in Arabic] *Syrian Observatory for Human Rights*, 18/6/2017.

'Al-Julani is Close to Faruq al-Shara: He Studied Jurisprudence at the Hands of a
Damascene Scholar in Mezze', [in Arabic] *Enab Baladi*, 28/7/2016.

'Al-Nusra Discharges One of its Founders, Social Media Accounts and Websites
Confuse him with Saleh Al-Hamawi', [in Arabic] *Zaman al-Wasl*, 16/7/2015.

'Al-Raqqa City is ISIS-Free', [in Arabic] *Syrian Observatory for Human Rights*,
21/9/2017.

'Around 585,000 People Have Been Killed Since the Start of the Syrian Revolution' [in Arabic] *The Syrian Observatory for Human Rights*, 4/1/2020.

'Fear of Islamic State Suicide Attacks Lingers in Raqqa', *Sky News*, 22/10/2017.

'Four Information about the Tabqa Military Airport in Raqqa', [in Arabic] *Enab
Baladi*, 27/3/2017.

'Video: The Islamic State Takes Control of 17th Division in the Countryside of
Raqqa', [in Arabic] *JBC News*, 27/7/2014.

'Including Missile Launchers and Artillery . . . The State Parades the 'Spoils' of the 121st Regiment', *Zaman al-Wasl*, [in Arabic] 28/7/2014.

'ISIS Announces Control over 17th Division', [in Arabic] *Zaman al-Wasl*, 25/7/2014.

'ISIS Appears Again in the Countryside of Raqqa, Attacking the Regime', [in Arabic] *Enab Baladi*, 14/1/2020.

'ISIS Controls the 93rd Brigade in Raqqa, Syria', [in Arabic] *Al Jazeera*, 7/8/2014.

'ISIS Controls the 17th Division in Raqqa', *Al Jazeera*, 25/7/2014.

'ISIS Ends the Battle for Tabqa Airport, the Regime Admits its Loss and the Shock Silences the Regime's Supporters', [in Arabic] *Zaman al-Wasl*, 24/11/2014.

'ISIS is Digging Underground Headquarters to Avoid Airstrikes in Raqqa', [in Arabic] *Syriadirect*, 5/10/2015.

'Islamic State Faces Endgame in Raqqa, says SDF', *Middle East Eye*, 20/9/2017.

'Study Shows Islamic State's Primary Opponent in Syria Is Government Forces', *Business Wire*, 19 April 2017.

'Interview: It will be a Fire that Burns the Cross and its People in Raqqah', *Rumiyah*, no. 12 (6/8/2017), p. 31.

'Jabhat Al-Nusra Expels the Leader 'Ous Al Sira' Fi Al Sham' Permanently', [in Arabic], *Arabi 21*, 15/7/2015.

'Jihadist Leader Abu Khaled Al-Suri was Killed in an Attack in Aleppo', [in Arabic] *BBC*, 24/2/2014.

'Jihadists Capture Key Base from Syrian Army', *The Daily Star Lebanon*, 8/8/2014.

'Kurdish Fighters Raise Flag of PKK Leader in Centre of Raqqa', *Middle East Eye*, 20/10/2017.

'Military Officials Reveal New Facts about the Fall of Tabqa Airport', [in Arabic] *Alsouria.Net*, 27/2/2016

'Pictures . . . Abu Khaled Al-Suri from Birth to Death in Aleppo', [in Arabic] *Taht Al-Mijhar*, 24/2/2014.

'17th Division Soldiers Dug their Graves with their Own Hands Before They were Executed in "Flames of War"', [in Arabic] *Zaman al-Wasl*, 22/6/2014.

'Syria: ISIS Controls the Headquarters of Brigade 93', [in Arabic] *The Syrian Observatory for Human Rights*, 8/8/2014.

'Syrian Activists: ISIS Stormed the Headquarters of the 93rd Army Brigade in Raqqa Causing Dozens of Deaths', [in Arabic] *CNN Arabic*, 7/8/2014

'Ten Operations for ISIS in Syria since the Beginning of 2020', [in Arabic] *Enab Baladi*, 15/1/2020.

'The Dispute between Al-Nusra and ISIS', [in Arabic] *Dar Eman TV*, 21/5/2016.

'US Gives Syria Intelligence on Jihadists: Sources', *The Daily Lebanon Star*, 26/8/2014.

'Video Leaked from the 93rd Brigade, Raqqa, after ISIS Took Control of it', [in Arabic] *Step News*, 7/8/2014.

'Weapons of the Islamic State: A Three-Year Investigation in Iraq and Syria', *Conflict Armament Research* (December 2017), pp. 183–4.

Chapter 4

Abdullah, Mohammed. Head of National Salvation Front Party. Interview by author. Istanbul, February 2016.

al-Arabi, Mohammad, 'Libya . . . ISIS Broadcasts Al-Baghdadi's Speeches in Sirte', [in Arabic] *Al Arabiya*, 13/2/2015.

Ashour, Omar, 'Between ISIS and a Failed State: The Saga of Libyan Islamists', in: Shadi Hamid and William McCants (ed.), *Rethinking Political Islam*, Oxford: Oxford University Press, 2017.

Belhaj, Abdul Hakim, former Commander (Emir) of the Libyan Islamic Fighting Group (LIFG). Interview by author. Doha, 27 January 2019.

Benotman, Noman, former Commander in the Libyan Islamic Fighting Group. Interview by author. Prague, 16 August 2019.

Besha, Abd al Aziz, 'The Islamic State Withdraws from Derna, Eastern Libya', [in Arabic] *Al Jazeera*, 20/4/2016.

Broder, Jonathan, 'Isis in Libya: How Muhammar Gaddafi's Anti-Aircraft Missiles are Falling into the Jihadists' Hands', *The Independent*, 11/3/2016.

Chang, Edward, 'The Harrier: The US Marine Corps Loves This Plane For 1 Big Reason', *The National Interest*, 29/3/2019.

Clausen, Christian, 'Providing Freedom from Terror: RPAs Help Reclaim Sirte', *Air Combat Command*, 1/8/2017.

Cruickshank, Paul, Nic Robertson, Tim Lister and Jomana Karadsheh, 'ISIS Comes to Libya', *CNN*, 18/11/2014.

Entous, Adam and Missy Ryan, 'In Libya, United States Lays Plans to Hunt Down Escaped Islamic State Fighters', *The Washington Post*, 11/11/2016.

Fitzgerald, Mary et al, 'A Quick Guide to Libya's Main Players', *European Council of Foreign Relations*, December 2016.

Fitzgerald, Mary, 'Libya's Rogue "War on Terror"', *Foreign Policy*, 5/6/2014.

Former Commander in *al-Bunyan al-Marsus* Operation. Interview by author. Istanbul, 12 November 2018.

Former commander in Ali Hasan al-Jaber Battalion (Bayda). Interview by author. Tunis, 23 September 2017.

el-Gomati, Anas, 'Haftar Rebranded Coups', *Sada*, 30/07/2019.

al-Haddad, Abdelwahab and Abu Bakr Al-Dharrat, 'The Confessions of "ISIS Dinosaurs" . . . The Princes' Driver Reveals the Leaders of the Organization and Smuggling Routes to Syria', [in Arabic] *Al-Araby al-Jadid*, 18/11/2018.

Hattem, Julian, 'CIA: Undaunted ISIS is Expanding, Focused on Attacking West', *The Hill*, 16/6/2016.

Lewis, Aidan, 'New Islamic State Leader in Libya Says Group "Stronger Every Day"'. *Reuters*, 10/3/2016.

Malsin, Jared, and Benoit Faucon, 'Islamic State's Deadly Return in Libya Imperils Oil Output', *Wall Street Journal*, 18/9/2018.

Mangush, Youssef, former Chief of Staff of the Libyan Armed Forces. Interview by author. Istanbul, March 2017.

Michael, Maggie, '16 Libyan Militiamen Killed in 2 IS Attacks near Sirte', *Fox News*, 17/6/2016.

Michael, Maggie, 'How a Libyan City Joined the Islamic State Group', *AP News*, 9/11/2014.

Michael, Maggie, 'ISIS Militants Retreat from Libya Bastion as Militias Advance', *Military Times*, 9/6/2016.

al-Muhajir, Abu Hudhayfa. Interview by Anonymous. *Rumiyah*, no. 4 (Hijri-dated Rabi' al-Awal 1438), p.10.

Nichols, Michelle, 'Islamic State in Libya Hampered by Lack of Fighters - U.N. Experts', *Reuters*, 1/12/2015.

Pengelly, Martin and Chris Stephen, 'Islamic State Leader in Libya "Killed in US Airstrike"', *The Guardian*, 14/11/2015.

Raghavan, Sudarsan, 'Even with US Airstrikes, a Struggle to Oust ISIS from Libyan Stronghold', *The Washington Post*, 7/8/ 2016.

Ryan, Miss, 'US Special Operations Troops Aiding Libyan Forces in Major Battle Against Islamic State', *The Washington Post*, 9/8/2018.

Ryan, Yasmine, 'Isis in Libya: Muammar Gaddafi's Soldiers are Back in the Country and Fighting Under the Black Flag of the "Islamic State"', *Independent*, 26/3/2015.

al-Saadi, Sami, former Commander in the Libyan Islamic Fighting Group (LIFG). Interview by author. Istanbul, 3 November 2016.

Schmitt, Eric and David D. Kirkpatrick, 'Islamic State Sprouting Limbs Beyond Its Base', *New York Times*, 14/2/2015.

al-Trabulsi, Seif al-Din, 'Daesh Bombing Kills 30 Soldiers Near Libya's Sirte', *Anadolu Agency*, 5/19/2016.

United Nation Security Council, 'Security Council ISIL (Da'esh) and al-Qaida Sanctions Committee Adds 12 Names to Its Sanctions List', *Press Release*, SC/12266, 29/2/2016.

al-Warfalli, Ayman, 'Libyan Security Forces Pushing Islamic State back from Vicinity of Oil Terminals', *Reuters*, 30/5/2016.

Wehrey, Frederic, *The Burning Shores: Inside the Battle for the New Libya*, New York: Farrar, Straus and Giroux, 2018.

Westcott, Tom, 'IS seizes Libya Airbase after Misrata Forces Pull out', *Middle East Eye*, 30/5/2015.

Westcott, Tom, 'No Ammo to Fight IS in Central Libya, Says Military', *Middle East Eye*, 5/6/2015.

Wilayat Barqa, 'Caliphate Soldiers Start a New War of Attrition Campaign', [in Arabic] *Al-Naba'*, no. 120, 22/2/2018, p. 6.

Wilayat Libya, 'Soldiers of the Caliphate in Libya Burn Down the Foreign Affairs Ministry', [in Arabic] al-*Naba'*, no. 162, 29/12/2018, p. 4.

Wilayat Libya, 'The Commander of the Soldiers of the Caliphate in Fezzan: We Shall Continue Storming Towns and Villages', [in Arabic] *Al-Naba'*, no. 182, 15/5/2019, p. 7.

Yacine, Hany, 'Would Libya Embrace ISIS?' [in Arabic] *Al-Ghad TV*, 5/10/2017.

al-Zawi, Soliman and David Kirkpatrick, 'Western Officials Alarmed as ISIS Expands Territory in Libya', *New York Times*, 31/5/2015.

Zelin, Aaron, 'The Others: Foreign Fighters in Libya', *Policy Notes of Washington Institute for Near East Policy*, no. 45 (2008), pp. 1–27.

Zelin, Aaron, *Your Sons are At Your Service*, New York: Columbian University Press, 2020.

'AFRICOM Concludes Operation Odyssey Lightning', *US Africa Command*, Press Release, 20/12/2016.

'An ISIS Military Parade in Sirte, Libya (Video and Photos)', [in Arabic] *Arabi 21*, 19/2/2015.

'Dismantling Armed Units of the Baath Party Apostates', [in Arabic] *al-Naba'*, no. 1, 17/10/2015, p. 9.

'Eulogy to Abu Nabil al-Anbari: Islamic State Leader in Libya', *aymennjawad. org*, 7/1/2016.

'*Fajr* Libya is Launching Air Strikes Against ISIS in Sirte', [in Arabic] *Al Jazeera Mubasher*, 24/4/2015.

'Former Libyan Official Ahmad Qadhaf al-Dam: I Support Isis, Which Should Have Been Established 50 Years Ago', *MEMRI TV*, The *Middle East Media Research Institute*, 17/1/2015.

'Gaddafi Soldiers are Fighting Under the Banner of the Islamic State', [in Arabic] *Al Quds Al Arabi*, 17/3/2015.

'Interview with Abdul Qadr al-Najdi, the Leader of IS in Libya', [in Arabic] *al-Naba'*, no. 21, 8/3/2016, pp. 8–9.

'Interview with Ahmed Gaddaf al-Dam, Gaddafi's cousin and one of the main leaders and former officials of his regime', *Dream TV*, Program 10 p.m., 17/1/2017.

'Interview with the Former Commander of the Abu Salim Martyrs Brigade: Salim Derby', [in Arabic] Program the Meeting of the Hour, *Libya Al-Ahrar TV*, 24/11/2019 (2013).

'Meeting of al-Bayda Tribal Elders with Ali Hassan al-Jaber Brigade's Commander in Fattaih-Derna', *YouTube*, 29/9/2015.

'"Mujahideen Shura Council" in Derna Declares War on "ISIS"', *Assabeel*, 11/6/2015.

'On the Killing of the Leader Salim Derby and the Developments in the City of Derna', [in Arabic] *YouTube*, 10/6/2015.

'Repentance of 42 Elements Working in the Ministry of Interior', [in Arabic] *Wilayat Tarabulus Media Office*, 14/2/2015.

'*al-Rimal al-Mutaharika*: Derna', [in Arabic], Documentary, 25/2/2018.

'*al-Rimal al-Mutaharika*: Sirte', [in Arabic] Documentary, 04/06/2017.

'Sirte Battle against ISIS is Entering its Final Stages', [in Arabic] *al-Araby al-Jadeed*, 12/8/2016.

'The Battle of Sirte: Hypotheses and Facts', [in Arabic] *Libya al-Khabar*, 11/12/2019.

'The Commander of the Operations Room of al-Fattaih Reveals the Name of ISIS Leader in the Area', [in Arabic] *al-Wasat*, 22/3/2016.

'The Islamic State is Showing its Force on the Streets of Sirte, Libya and the University Closes its Doors', [in Arabic] *France 24*, 19/2/2015.

'The Islamic State Media Office in Tripoli: The Battle of Sheikh Abu Anas al-Libi', [in Arabic] *justpaste.it*, 5/4/2015.

'Three British ISIS Members Killed by SBS Soldiers in Sirte', *Libyan Express*, 4/7/2016.

'United Nations Security Council Report', *United Nations Security Council*, S/2019/570, 15/7/2019.

'IS Militants Claim Deadly Bombings in Libya', *VOA News*, 20/2/2015.

'ISIS Admits the Loss of Derna', [in Arabic] *Afrigatenews.net*, 12/7/2015.

'ISIS Admits the Loss of Derna on Video', [in Arabic] *Arabi 21*, 12/7/2015.

'Libyans "Oust" So-Called IS from Sirte Headquarters', *BBC*, 10/8/2016.

'US Airstrikes in Support of the GNA, October 13', *US Africa Command Press Release*, 14/10/2016.

'Statement from Pentagon Press Secretary Peter Cook on Nov. 13 airstrike in Libya', *US Department of Defense*, Immediate Release, 7/12/2015.

Chapter 5

Abdul Azim, Ahmed, 'Major General Ahmed Wasfi to "El-Watan": The Military Operations in Sinai are Over', [in Arabic] *El- Watan*, 12/10/2013.

Abu Rumman, Mohamed and Hasan Abu Haniya, *The Islamic State: The Sunni Crisis and the Struggle over Global Jihadism*, [in Arabic] Amman: Friedrich-Ebert, 2015.

Akeel, Yehia, Former MP of North Sinai Governorate. Interview by author, 6 April 2015.

al-Anani, Mahmoud, 'Desert Flames: Sinai Province Confirms "We Are Here"', [in Arabic] *Noon Post*, 3/8/2016.

Anonymous Member of *Ansar Bayt al-Maqdis*. Interview by Mona al-Zamlout entitled 'al-Sisi is an Apostate and We Are Continuing to Fight', [in Arabic], *Al Jazeera*, 28/5/2014.

Arreguin-Toft, Ivan, *How the Weak Win Wars?*, Cambridge: Cambridge University Press, 2005.

Ashour, Omar, 'Collusion to Collision: Islamist-Military Relations in Egypt', *Brookings Papers*, no. 14 (March 2015), pp. 1–44.

Ashour, Omar, *The De-Radicalization of Jihadists: Transforming Armed Islamist Movements*, New York, London: Routledge, 2009.

Ashour, Omar, 'A Not-So-Inevitable Defeat: Combat Performance on the Egyptian Front', [in Arabic] in: Ahmed Hussein (ed.), *The June 1967 War: Trajectories and Ramifications*, Doha: Arab Centre for Research and Policy Studies, 2020.

Atef, A, 'The Military Spokesperson: 241 from the Armed Elements were Killed in Sinai in Five Days', [in Arabic] *Akhbarak*, 6/7/2015.

al-Baghdadi, Abdul Latif, *Memoirs of Abdul Latif al-Baghdadi – Part One*. [in Arabic] Cairo: Al-Maktab al-Masri, 1977.

Barayez, Abdul Fattah, 'This Land is Their Land: Egypt's Military', *Jadaliyya*, 25/1/2016.

Bilal, Ahmed, 'New Surprises in the Sheikh Zuweid Battle', [in Arabic] *Almasry Alyoum*, 14/7/2015.

Chaitani, Youssef, Omar Ashour and Vito Intini, 'An Overview of the Arab Security Sector amidst Political Transitions: Reflections on Legacies, Functions and Perceptions', *United Nations' Economic and Social Commission for West Africa (UN-ESCWA)*, New York: United Nations Publications, July 2013.

De Atkine, Norvell, 'Why Arabs Lose Wars?', *Middle East Quarterly* (December 1999), pp. 1–5.

Egyptian Observatory of Rights and Freedoms (EORF), 'Sinai . . . Two Years of Crimes', [in Arabic] *Klmty*, 15/6/2015.

Former Islamist Detainee. Interview by the author. Cairo, September 2012.

Former Major in Central Security Forces who served in Sinai. Interview by author, August 2016.

Former Major in Central Security Forces who served in Sinai. Interview by author, December 2016.

Former Major-General in the Egyptian Armed Forces. Interview by author. London, 29 September 2015.

Former Major-General in the Egyptian Military Intelligence. Interview by author, 6 November 2016.

Former Prison Warden of the 'Scorpion' and Liman Tora Prisons. Interview by the author. Cairo, September 2012.

Galula, David, *Counterinsurgency Warfare: Theory and Practice*, London: PSI, 2006

Gray, G, 'Internal Security in Sinai', *WikiLeaks*, WikiLeak Cable no. 05CAIR01978.

Hassan, Ahmed, 'One of Us: The Militant Egypt's Army Fears Most', *Reuters*, 16/10/2015.

Human Rights Watch, 'All According to Plan: The Raba'a Massacre', Human Rights Watch Reports, 12/8/2014.

Jumaa, Ahmed, '10 Pieces of Information about Abu Hajar al-Hashemi, the Master Planner of the Terrorist Attack on the al-Rawda Mosque', [in Arabic] *Youm 7*, 30/11/2017.

Local Sinaian Journalist. Interview by author via phone, August 2015.

Major-General in the Egyptian Ministry of Interior. Interview by the author. Cairo, 13 April 2013.

al-Marzouki, Eid, 'The Tragedy of Sinai in the Story of Youssef Zera'i', [in Arabic] *al-Araby al-Jadid*, 6/3/2015, p. 6.

al-Marzouki, Eid, Sinaian Bedouin activist. Interview by author, Doha. October 2015.

Media Office of Sinai Province, 'The Annual Harvest of Military Operations for the Year 1436', [in Arabic] 12/11/ 2015.

Media Office of Sinai Province, 'The Annual Harvest of Military Operations for the Year 1437' [in Arabic], 10/11/2016.

al-Naquib, Amrun, 'Sources for 24: The Iraqi "Abu Hajar Al Hashemi" is the Master Planner for the Massacre of the Rawda Mosque', [in Arabic] *24*, 20/11/2017.

al-Naquib, Amrun, 'Kamal Allam: From a Popular Singer to the Emir of al-
Tawhid wa al-Jihad', [in Arabic] 24, 28/12/2017.
Pollack, Kenneth, Arabs at War, Lincoln: Nebraska University Press, 2002.
Porch, Douglas, Counterinsurgency: Exposing the Myths of the News Way of War,
Cambridge: Cambridge University Press, 2013.
Sayigh, Yezid, 'Above the State: The Officers Republic of Egypt', Carnegie
Papers (August 2012), pp. 1–28.
Schmidt, Dana, Yemen: The Unknown War, New York: Holt, Rinehart and
Winston, 1968.
Sharp, Jeremy, 'Egypt: Background and U.S. Relations', Congressional Services
Report, 5/6/ 2014.
al-Tahwid wa al-Jihad – Bayt al-Maqdis. This is Our Creed, [in Arabic], Site of
publication unknown: no publisher, 2009.
Tantawy, Mohamed, 'The Harvest of 11 Days', [in Arabic] Al-Youm al-Sabi',
18/9/2015.
Turki, Sayed and Wael Mamdouh, 'How did Sheikh Zuweid Escape the Fate of
Mosul and Ramadi? (News Analysis)', [in Arabic] Almasry Alyoum, 8/7/2015.
al-Zamlout, Mona, 'Ansar Bayt al-Maqdis: We are not Affiliated with the Muslim
Brotherhood or with al-Qaida', [in Arabic] Aljazeera, 28/5/2014.
'A Video Clip Ignites a Controversy Over the Role of "Ali Mamlouk" in the
Assassination of an Egyptian Officer', [in Arabic] Orient Net, 30/10/2016.
'Bayt al-Maqdis Publishes a Video of a Helicopter Shot Down in the Sinai', [in
Arabic] Al Arabiya, 27/1/2014.
'Deaths in the Egyptian Army', [in Arabic] Al-Jazeera, 12/9/2015.
'Egypt's Sinai Rocked by Wave of Deadly Attacks', BBC, 1/7/2015.
'Evening Window: An Explosion of State Security and Losses on the Egyptian
Stock Exchange', [in Arabic] Al Jazeera News, 20/8/2015.
'For the Second Day, Funerals of Youth Killed by the Security Turns into
Demonstrations', [in Arabic] Al-Araby Al-Jadid, 16/1/2017, p. 1.
'Interview with the Emir of the Caliphate Soldiers in Egypt', Al-Naba',
Hijri-dated 8 Sha'ban 1438, p. 8.
'Look for Another Homeland', Human Rights Watch, 22/9/2015.
'Sinai Province: Egypt's Most Dangerous Group', BBC, 12/5/2016.
'The Military Spokesperson: Death of 170 Terrorists in Sinai', [in Arabic] MBC,
1/3/2015.

Chapter 6

Ashour, Omar, 'The Enigma of Endurance and Expansion', [in Arabic] Policy
Studies, Doha: al-Jazeera Centre for Studies, March 2016.
Bajis, Mu'ayad, 'For the First time . . . the State Organisation Reveals its Structure
in Details', [in Arabic] Arabi 21, 06/07/2016.
De Atkine, Norvell, 'Why Arabs Lose Wars?', Middle East Quarterly (December
1999), pp. 1–5.
Former Commander in al-Bunyan al-Marsus Operation. Interview conducted by
author, Istanbul, 12 November 2018.

Al-Husan, Muhammad. Commander in the 166th Battalion. Interview in: '*Al-Rimal al-Mutaharika*: Sirte', [in Arabic], Documentary, 25/2/2018.

Kaaman, Hugo, 'Factories of Destruction', *Hugo Kaaman Open Source on SVBIEDs*, 31/03/2020.

Montgomery, Bernard, *A History of Warfare*, London: Collins, 1968.

al-Mustafa, Hamzeh, 'The Islamic State in Syria: Origins and the Environment', [in Arabic] in: Azmi Bishara (ed.), *The State Organisation (Acronym ISIS): Foundation, Discourse and Practice* – Vol. 2 [in Arabic], Doha: the Arab Centre for Research and Policy Studies, 2018.

Pollack, Kenneth, *Arabs at War*, Lincoln: Nebraska University Press, 2002.

Shamir, Eitan, *Transforming Command: The Pursuit of Mission Command in the U.S., British, and Israeli Armies*, Stanford: Standard University Press, 2011.

al-Tamimi, Aymenn, 'Dissent in the Islamic State: Abu al-Faruq al-Masri's Message on the *Menhaj*', *Combating Terrorism Center*, 31/10/2016.

Thoma, Dominik, *Moltke Meets Confucius: The Possibility of Mission Command in China*, Tectum Verlag, 2016.

Wider, Werner, '*Auftragstaktik* and *Innere Fuhrung*: Trademarks of German Leadership', *Military Review* (September–October 2002), pp. 3–9.

'Interview with Abu Mus'ab al-Zarqawi', [Arabic] *Minbar al-Tawhid wa al-Jihad* (Hijri-year 1427 or 2006).

'Interview: It will be a Fire that Burns the Cross and its People in Raqqah', *Rumiyah*, issue no. 12 (06/08/2017), p. 33.

'Iraq is in "Daily Battles" against the State Organisation in Kirkuk', [in Arabic] *al-Arab*, 19/04/2020.

'More than 1,000 Operations in 4 Months Kills 135 Terrorists', [in Arabic] *Iraqi24*, 23/04/2020.

'President Trump on Syria: I'm not Siding with Anybody', *C-Span*, 7/10/2019.

'Soldiers' Harvest', *al-Naba'*, issue no. 224, Hijri-dated 10 Rajab 1441, p. 12.

Index

territorial defeats and horizontal
escalation, 195–7
unique aspects of combat
performance, 202–9
weapons production, 41–3
Islamic State Ways of Warfare
Database (ISWD), 25, 129,
139, 141, 143, *144*, 149n26
Islamic Youth Shura Council
(IYSC), 126, 128, 129, 131

Jabhat al-Nusra li Ahl al-Sham (JN),
2, 79, 83–4, 87, 88, 90–2
Jaysh al-Islam (Army of Islam), 90,
113n32
al-Jazrawa, Abu Omar, 198
Jihadism, 18, 80, 111n13, 153n64,
164–5, 167
Johnston, Patrick, 11
Jones, Seth, 5, 9, 11–12
al-Juburi, Maysara, 80, 111n12
al-Julani, Abu Muhammad, 79–80,
81, 83–4, 110n5

Kalyvas, Stathis N., 8, 10
Keane, Gen Jack, 38
Keefer, Philip, 7
Khalid, Abu, 177
Kharijites, 151n41
Khattab, Anas, 80
al-Khlifawi, Samir, 45
Kilcullen, David, 6, 11
knowledge transfer, 83, 87,
106, 128, 129–30, 145–6,
166–7, 197, 198, 202–3,
205, 210

Laitin, David D., 8
Lansdale, Edward G., 5–6
Latakia, 24
Lawrence, Thomas Edward, of
Arabia, 6, 7
Libicki, Martin C., 4–5, 8–9

Libya, 2, 126–59, 196
battlefields analysis, 140–5, *142,
144*
al-Bunyan al-Marsus operation, 138
casualties, 137, 139
Derna battlefront June 2015–
April 2016, 130–4, 141–2, *142*
encircle and choke tactics,
131–2, 141
Government of National Accord
(GNA), 134, 138
IS beginnings in Libya, 126–7
IS insurgency, future of, 145–8
IS military build-up, overview
of, 127–30
al-Nufaliya battle, 136
Sirte battlefront March
2015–December 2016,
135–40
weaponry, 129–30, 132, 143,
144–5, 146
Libya Dawn, 136–7
Libyan Islamic Fighting Group
(LIFG), 80, 111n13
Libyan National Army (LNA), 131,
134, 147, 150n35
Liwa' al-Tawhid (Monotheism
Brigade), 84
Loss Exchange Ratio (LER), 3
Loss of Accuracy Gradient (LAG),
8, 15, 51–2
Loss of Strength Gradient (LSG),
8, 15
Luqman, Abu, 199
Lyall, Jason, 4, 9, 11

Macaulay, Neil, 7
McChrystal, Gen Stanley, 6
McColl, Robert, 5, 8
Mahdi, Adel Abdul, 195
man-portable air-defence systems
(MANPADS), 9, 42, 59,
69n36, 86, 157n126, 181

al-Qaida (AQ), 10, 27, 81,
    167, 168
al-Quraishi, Abu Ibrahim
    al-Hashami, 195

Ramadi, 2, 39, 41, 199, 201, 203,
    204, 208
  battlefront May 2015–February
    2016, 54–7
  tunnels and other tactics, 62–4, 63
Raqqa, 1, 23, 26, 81–2, 199, 200–1,
    203, 204, 208
  armoured units, 105–6
  battlefronts analysis, 99–106, 100
  defences and counteroffensives,
    94–9
  events April 2013–October
    2017, 88–99
  guerrillas, suicide guerrillas and
    snipers, 102–5
  intelligence, 106
  IS expansion, 89–92
  opposing the Assad regime, 92–4
  revenge attacks, 108
  SVBIEDS and other
    IED-intensive tactics, 100–2
recruitment, 83–4, 88, 128, 135,
    171–2, 183, 198–9
research design, 23–5
research findings, 197–210
  non-unique aspects of IS combat
    performance, 198–201
  operational level, 202–4
  strategic shifts and failures,
    204–5
  tactical innovations and
    successes, 205–9, 206
  unique aspects of IS combat
    performance, 202–9
Rif Dimashq, 1
rockets, 57, 59, 85, 170, 179, 181
Rumiyah magazine, 25, 106, 138–9,
    143

Sahd, Col Salih, 134
Salafism, 18–19
Salahuddin, 1, 24, 41
al-Salih, Mustafa, 80, 111n11
Salih, Abdul Qadir, 84
sanctuary, 9
SCCLC modus operandi, 47–8,
    51, 55, 59, 66, 99, 109, 147,
    184, 203
Schutte, Sebastian, 8
Seig, Hans, 11
al-Sha'iri, Hasan al-Saliheen, 128
Sheikh Zuweid, 23, 36n114, 160,
    176–7, 182–4, 199, 200, 203
al-Shishani, Abu Omar, 198
Shukri, Col Ismail, 133, 140, 144,
    145
Sinai Observatory (SO), 172
Sinai Province (SP), 23, 26,
    160–94, 196
  absorption by the Islamic State,
    166–8
  attacks, 160–1, 169–71
  attacks on tourists, 163–4
  battlefronts analysis, 178–85,
    179, 180
  conspiracy theories, 168–9
  counterinsurgency operations,
    165–6, 172–5
  endurance and survival strategy,
    184
  environmental factors affecting
    success, 175–6
  future of SP, 185–6
  geographical factors, 162
  local support problems, 183
  military build-up and survival
    strategies, 161–2
  military capacity, 169–72
  military development of,
    162–3
  military endurance, reasons for,
    168–76

CPSIA information can be obtained
at www.ICGtesting.com
Printed in the USA
JSHW021704080521
14528JS00004B/91